Inverse and Ill Posed Problems Series

Inverse and Ill-Posed Sources Problems

Also available in the Inverse and Ill-Posed Problems Series:

Methods for Solution of Nonlinear Operator Equations
V.P. Tanana

Nonclassical and Inverse Problems for Pseudoparabolic Equations
A. Asanov and E.R. Atamanov

Methods for Solving Operator Equations
V.P. Tanana

Formulas in Inverse and Ill-Posed Problems
Yu.E. Anikonov

Inverse Logarithmic Potential Problem
V.G. Cherednichenko

Multidimensional Inverse and Ill-Posed Problems for Differential Equations
Yu.E. Anikonov

Ill-Posed Problems with A Priori Information
V.V. Vasin and A.L. Ageev

Integral Geometry of Tensor Fields
V.A. Sharafutdinov

Inverse Problems for Maxwell's Equations
V.G. Romanov and S.I. Kabanikhin

Other related books from VSP:

Inverse Problems of Mathematical Physics
V.G. Romanov

Inverse Sturm-Liouville Problems
B.M. Levitan

Related Journal:

Journal of Inverse and Ill-Posed Problems
Editor-in-Chief: M.M. Lavrent'ev

INVERSE AND ILL-POSED PROBLEMS SERIES

Inverse and Ill-Posed Sources Problems

Yu.E. Anikonov, B.A. Bubnov and G.N. Erokhin

///VSP///
Utrecht, The Netherlands, 1997

VSP BV
P.O. Box 346
3700 AH Zeist
The Netherlands

First published in 1997

ISBN 90-6764-273-8

Printed in The Netherlands by Ridderprint bv, Ridderkerk.

Contents

Introduction **3**

1 The formulas and constructive methods for solving inverse problems of source determination. **5**

§1.1 The Meyer transformation in the inverse problem with axial symmetry. 5
§1.2 The structure approach to inverse problem solving. . 11
§1.3 Determining the index and the right side of the pseudodifferential equation. 26
§1.4 Symmetry in inverse problems of source determination. . 31
§1.5 Inversion formulas for the wave equation. Setting of the problem. 34
§1.6 Conjugated equations in control problems of nonstationary geophysical fields. 48
§1.7 Inverse problem of source determination as the problem of statistical evaluation of parameters of the mathematical model. 53
§1.8 Formulas in multidimensional inverse problems for evolutionary equations. 62

2 Correctness of certain inverse problems. **69**

§2.1 The Cauchy problem for elliptic and ultrahyperbolic equations in the classes of finite smoothness functions and inverse problems. 69
§2.2 Inverse problems for multidimensional hyperbolic equations. 85
§2.3 Control problems and inverse problems. 116

§2.4 Inverse problems for nonlinear equations 121
§2.5 Inverse problems of transfer theory. 134
§2.6 Geometric problems. 141

3 Numerical solution of direct and inverse problems. 159
§3.1 Mathematical simulation on the basis of the seismic moment tensor theory. 159
§3.2 Determining the seismic moment tensor of the seismic centre. 183
§3.3 The integral transformations method in inverse problems of source determination. 195
§3.4 The problem of distance quality control of an airplane landing. 222

Bibliography 231

Introduction

In this monograph, the authors have considered the different settings of inverse problems of mathematical physics, both in linear and in nonlinear cases. Most attention has been concentrated on unique solvability and on the search for constructive methods. As a rule, the problems considered here are multidimensional. Effective methods for source recovery are described for sources, of different nature. In studies of these problems the following stages may be distinguished.

1. Analytic representation of physical sources.

2. Mathematical simulation of physical fields induced by sources of the given type.

3. Comparison of real and theoretical information. Correction of mathematical models of sources.

4. Theoretical investigation of the inverse problems and construction of effective numerical algorithms.

5. Development of a program complex for problem solving during real time.

The authors hope that the suggested methods and results will be used for development both of theoretical and practical problems of simulation and determining the sources on the basis of multidimensional inverse problems of mathematical physics.

This monograph, together with previous monographs of Anikonov Yu. E., "Multidimensional Inverse and Ill-Posed Problems for Differential Equations" and "Formulas in Inverse and Ill-Posed Problems", propose a wide constructive collection of investigation methods in this area.

The main results of this book belong to the authors.

This work was suppoted by the Russian Foundation for Fundamental Research, grant number 96-01-01558.

Chapter 1

The formulas and constructive methods for solving inverse problems of source determination.

In this chapter we describe various methods for solving inverse problems of source determination on the basis of integral transformations, the structure method and optimization methods. The main goal is creation of constructive solution methods.

§1.1 The Meyer transformation in the inverse problem with axial symmetry.

In the cylindrical coordinate system consider the wave equation with damping

$$\frac{\partial^2 u}{\partial z^2} + \frac{1}{r}\frac{\partial}{\partial r}(ru) - \left(\frac{1}{c}\frac{\partial}{\partial t} + i\alpha\right)^2 u = g(r)f(z,t), \tag{1.1}$$

where $u = u(r,z,t)$; $z, t \in R^1$; $r \in R^1_+$; $c > 0, \alpha > 0$, $R^1_+ = \{r, r \geq 0\}$.

It is required to determine a function $f(z,t)$, given the function $g(r)$ and the solution $u(r,z,t)$ trace in the line

$$u(r,z,t)\Big|_{z=0} = u_0(r,t). \tag{1.2}$$

Such setting is common for certain inverse problems of acoustics and seismology where it is required to determine the behavior of axially symmetric radiation sources (which may be moving) by the wave field measured in the earth surface.

Let $S'(R^n)$ be the space of generalized functions of slow growth (see, for example, Vasiljev, 1981); and let $\mathcal{K}'_{0,a}$ (Brychkov and Prudnikov, 1977; Zemanyan, 1974) be the space conjugate to $\mathcal{K}_{0,\alpha}$, consisting of all complex valued infinitely differentiable functions $\varphi(r)$, such that the functionals

$$\rho^0_{a,k} = \sup_{0<r<\infty}\left|\frac{e^{ar}S_0^k\varphi(r)}{\sqrt{r}h(r)}\right|$$

are finite for $k=0,1,\ldots$. Here

$$h(r) = \begin{cases} \ln r, & 0<r<e^{-1} \\ -1, & e^{-1}\le r<\infty, \end{cases}$$

$$s_0^k = \left(\frac{1}{\sqrt{r}}\frac{\partial}{\partial r}\left(\frac{1}{\sqrt{r}}\right)\right)^k.$$

It is known (Brychkov and Prudnikov, 1977) that $\mathcal{K}'_{0,a}\subset D'(R^1_+)$, where D' is the space of generalized functions (Vasiljev, 1981). As the functions $\sqrt{\eta^r}K_0(\eta r)$, where K_0 is the McDonal'd function of zero order, $\mathrm{Re}(\eta)>a$, $\eta\neq 0$, belongs to the basis set $\mathcal{K}_{0,a}$, we introduce for generalized functions from $\mathcal{K}_{0,a}$ the integral Meyer transformation or the K-transformation, which has the form (Brychkov and Prudnikov, 1977)

$$\hat f(\eta) = \big(f(r), \sqrt{\eta r}K_0(\eta r)\big). \tag{1.3}$$

Denote by $\mathcal{M}_0$ the set of such functions $\psi(r,z,t)\in D'(R^1_+\times R^2)$ that $\psi\in\mathcal{K}'_{0,0}$ by r and $\psi\in S'(R^2)$ by z and t.

Theorem 1.1. Let a function $f(z,t)$ belonging to $S'(R^2)$, be even by z, and suppose that it has a compact support without the point $z = 0$. Let $g'(r) = g(r)\sqrt{r} \in \mathcal{K}'_{0,0}$ and $\hat{g}(\eta) \neq 0$. Then the inverse problem of determining $f(z,t)$ by the trace $u_0(r,t)$ of solution $u(r,z,t) \in \mathcal{M}_0$ has a unique solution satisfying the relation

$$\left(\hat{f}(z,\omega),\varphi\right) = \left(\frac{1}{2\pi i}\int\limits_{\sigma-i\infty}^{\sigma+i\infty}\int\limits_{0}^{\infty}\hat{u}_0(r,\omega)K_0\left(r\sqrt{\bar{\omega}^2-q^2}\right)\right.$$

$$\left.\times\frac{(\bar{\omega}^2-q^2)^{1/2}}{\hat{g}\left(\sqrt{\bar{\omega}^2-q^2}\right)}e^{zq}rdrdq,\varphi\right) \quad \forall\varphi\in S'(R^2). \tag{1.4}$$

Here $\hat{f}(z,\omega)$ and $\hat{u}_0(r,\omega)$ are the Fourier transformations of $f(z,t)$, $u_0(r,t)$; $\bar{\omega} = \alpha + i\frac{\omega}{c}, \alpha > 0$, $\sigma > 0$, $\mathrm{Re}\left(\sqrt{\bar{\omega}^2-q^2}\right) > 0$. The function $\hat{g}'(\eta)$ is the K-transformation of $g'(r) = g(r)\sqrt{r}$ determined by (1.3).

Proof. Applying to (1.1) the Fourier transformation by time and replacing the functions

$$u'(r,z,\omega) = u(r,z,\omega)\sqrt{r}, \qquad g'(r) = g(r)\sqrt{r},$$

we have

$$\frac{\partial^2 u'}{\partial z^2} + \frac{1}{\sqrt{r}}\frac{\partial}{\partial r}\left(r\frac{u'}{\sqrt{r}}\right) - \left(\alpha + i\frac{\omega}{c}\right)^2 u' = g'(r)\hat{f}(z,\omega). \tag{1.5}$$

Applying to (1.5) the K-transformation (1.3) and taking into account its properties, we obtain instead of (1.5)

$$\frac{d^2\hat{u}'}{dz^2} - \left(\bar{\omega}^2 - \eta^2\right)\hat{u}' = \hat{f}(z,\omega)\hat{g}'(\eta), \tag{1.6}$$

where $\hat{u}' = \hat{u}'(z,\eta,\omega)$, $\bar{\omega} = \alpha + i\frac{\omega}{c}$, $\alpha > 0$, $\mathrm{Re}(\eta) > 0$.

The solution of (1.6) in the space of generalized functions of slow growth by z is as follows

$$\bar{u}'(z,\eta,\omega) = \hat{g}'(\eta)\int\limits_{-\infty}^{\infty} f(\tau,\omega)\frac{e^{-|z-\tau|\sqrt{\bar{\omega}^2-\eta^2}}}{\sqrt{\bar{\omega}^2-\eta^2}}d\tau. \tag{1.7}$$

As is seen from (1.7), we have chosen the positive branch $\mathrm{Re}\left(\sqrt{\bar{\omega}^2-\eta^2}\right)$. As the support of $\hat{f}(z,\omega)$ is separated from zero, then the trace (1.2) of (1.1) solution is correctly determined. Since $\hat{f}(z,\omega)$ is even by z, after substitution $\bar{\omega}^2-\eta^2=g^2$, (1.7) takes the form of a one-sided Laplace transformation for the function with a finite support

$$\int_0^\infty f(z,\omega)e^{-zq}dz = \frac{2\bar{u}\left(0,\sqrt{\bar{\omega}^2-\eta^2},\omega\right)}{\hat{g}'\left(\sqrt{\bar{\omega}^2-\eta^2}\right)}q. \tag{1.8}$$

Expressing $\hat{u}'(0,\eta,\omega)$ *via* $u_0(r,\omega)$ by the formula

$$\hat{u}'(0,\eta,\omega)=\left(\hat{u}_0(r,\omega)\sqrt{r},\sqrt{\eta r}K_0(\eta r)\right)$$

and formally inversing (1.8), we obtain the desired representation (1.4).

Theorem 1.2. Let an even by z function $f(z,t)$ belong to $S'(R^2)$, $\{z=0\}\cap \mathrm{supp} f(z,t)=\{0\}$, and suppose $g'(r)\subset \mathcal{K}^1_{0,0}$ has the form

$$g'(r)=\frac{\delta(r)}{\sqrt{r}\ln r}. \tag{1.9}$$

Then the solution to inverse problem (1.1)-(1.2) is determined by the formulas

$$f(z,\omega)=(f_1,\varphi)+(f_2,\varphi)+(f_\alpha,\varphi), \tag{1.10}$$

$$f_1(z,\omega)=\frac{1}{2}\frac{d}{dz}\int_0^\infty \hat{u}_0(r,\omega)\frac{e^{-\bar{\omega}\sqrt{r^2+z^2}}}{\sqrt{r^2+z^2}}rdr, \tag{1.11}$$

$$f_2(z,\omega)=\frac{1}{2}\frac{d}{dz}\int_{i\frac{\omega}{c}}^{i\infty}\int_0^\infty \hat{u}_0(r,\omega)J_0\left(r\sqrt{u^2-\bar{\omega}^2}\right)e^{zu}rdrdu, \tag{1.12}$$

$$f_\alpha(z,\omega)=\frac{1}{2}\frac{d}{dz}\int_{i\frac{\omega}{c}}^{\alpha+i\frac{\omega}{c}}\int_0^\infty \hat{u}_0(r,\omega)J_0\left(r\sqrt{u^2-\bar{\omega}^2}\right)e^{zu}rdrdu. \tag{1.13}$$

Here $\bar{\omega}=\alpha+i\frac{\omega}{c}$, $\alpha>0$, $\mathrm{Im}\left(\sqrt{u^2-\bar{\omega}^2}\right)>0$.

Proof. We shall consider the case when $g'(r)$ is the analog of the Dirac function from the basis functions space $\varphi(r) \in \mathcal{K}_{0,0}$; i.e., $g'(r)$ is such that

$$\left(g'(r), \psi(r)\sqrt{r}\ln r\right) = \psi(0).$$

Therefore, the test function $\sqrt{\eta r}K_0(\eta r) \in \mathcal{K}_{0,0}$, $\mathrm{Re}(\eta) > 0$ will satisfy the relation

$$\hat{g}'(r) = \left(g'(r), \sqrt{\eta r}K_0(\eta r)\right) = \sqrt{\eta}. \tag{1.14}$$

Using the δ-function symbolism, $g'(r)$ may be represented in the form (1.9); therefore

$$g(r) = \frac{\delta r}{r \ln r}.$$

Substituting (1.14) into (1.4), we obtain

$$\left(\hat{f}(z,\omega), \varphi\right) = \left(\frac{1}{2\pi i}\frac{d}{dz}\int\limits_{\sigma-i\infty}^{\sigma+i\infty}\int\limits_{0}^{\infty} \hat{u}_0(r,\omega)K_0\left(r\sqrt{\bar{\omega}^2 - q^2}\right)e^{zq}r\,dr\,dq, \varphi\right). \tag{1.15}$$

As $\hat{f}(z,\omega)$is finite in z, and its Laplace transformation is an entire analytic function, the integration contour may be deformed as is shown in Figure 1.1.

Here, $q = \alpha + i\frac{\omega}{c}$ is the branch point. In this case, in the upper cut shore γ_u and in the half-interval γ_l we have (according to Golubyatnikov, 1982)

$$K_0(ze^{i\pi}) = K_0(z) - i\pi I_0(z),$$

i.e., two functions $K_0(z)$ and $I_0(z)$, where $I_0(z)$ is the modified Bessel function of zero order. Thus, integrating along the imaginary axis gives us two terms

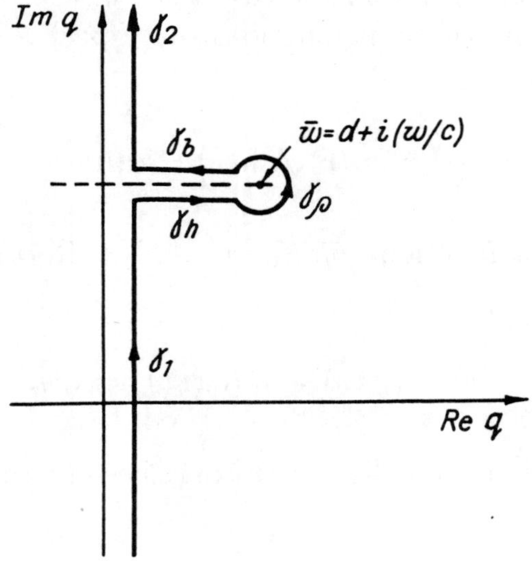

Figure 1.1. The integration countour.

$$\hat{f}_1(z,\omega) = \frac{1}{\pi}\frac{d}{dz}\int_0^\infty\int_0^\infty \hat{u}_0(r,\omega)K_0\left(r\sqrt{\bar{\omega}^2+q^2}\right) rdr\cos zqdq,$$

$$\hat{f}_2(z,\omega) = -\frac{1}{2}\frac{d}{dz}\int_{i\frac{\omega}{c}}^{i\infty}\int_0^\infty \hat{u}_0(r,\omega)I_0\left(r\sqrt{\bar{\omega}^2-q^2}\right) e^{zq}rdrdq.$$

Changing the integration limits in the first term and calculating the interior integral by means of (43) from Bejtman and Erdein (1969), we obtain the representation of the form (1.13). It may be shown that the sum of integrals by γ_u and γ_l gives

$$\hat{f}_\alpha(z,\omega) = \frac{1}{2}\frac{d}{dz}\int_0^{\alpha+i\frac{\omega}{c}}\int_0^\infty \hat{u}_0(r,\omega)I_0\left(r\sqrt{\bar{\omega}^2-q^2}\right) e^{zq}rdrdq.$$

The integral along the circle c_ρ for $\rho\to\infty$ tends to zero. Really, substituting $q = \bar{\omega} + \rho e^{i\varphi}$ and changing the integration order for the interior integral we have

$$R_{c_\rho}(r,z,\omega) = \int\limits_{c_\rho} K_0\left(r\sqrt{\bar{\omega}^2 - q^2}\right) e^{zq} dq$$

$$= i\rho \int\limits_{-\pi}^{+\pi} K_0\left(r\sqrt{\rho e^{i(\varphi+\pi)}(\rho e^{i\varphi} + 2\bar{\omega})}\right) e^{\rho z(\cos\varphi + i\sin\varphi)} e^{i\varphi} d\varphi.$$

Hence, since $K_0(\tau) \to \ln\tau$ as $\tau \to 0$, it is easy to show that $|R_{c_\rho}| \to 0$ as $\rho \to 0$.

Remark. Apply to the obtained representation of the source function (1.11) the inverse Fourier transformation in time. After changing the order of integration, we obtain the formula for $f(z,t)$

$$\left(f(z,t),\varphi\right) = \left(\frac{1}{2}\frac{d}{dz}\int\limits_0^\infty u_0\left(r, t + \frac{\sqrt{r^2+z^2}}{c}\right)\frac{e^{\alpha\sqrt{r^2+z^2}}}{\sqrt{r^2+z^2}} r dr, \varphi\right)$$

$$+\left(f_2(z,t),\varphi\right) + \left(f_\alpha(z,t),\varphi\right), \quad \forall\varphi \in S(R^2).$$

The obtained solution is inversion of the "direct" formula, which for our case is as follows

$$u_0(r,t) = 2\int\limits_0^\infty f\left(z, t - \frac{\sqrt{r^2+z^2}}{c}\right)\frac{e^{-\alpha\sqrt{r^2+z^2}}}{\sqrt{r^2+z^2}} dz.$$

The first term in the obtained formula describes the so-called "summing along the hodograph" (Korlyuch, 1985); the second term corresponds to accounting inhomogeneous waves (Trev, 1984).

§1.2 The structure approach to inverse problem solving.

The need to consider, from the uniform point of view, more and more complicated and correlated processes has led to the structure scheme method for distributed systems (Butkovskii, 1985). By means of this method, the complicated interactions of various fields may be represented. The basic notion of this method is that of a distributed block, where the entrance and exit are selected; and the input signal uniquely transforms into an output signal. The collection of such blocks with a description of its combining rules (parallel, sequential, with closing by an inverse relation and so on) gives a visual scheme of solving direct problem by steps. The inverse problem for a linear differential equation, from the formal point of view, is either to determine the transmitting block function (determining differential operator), or to determine the input signal (the right side of the equation) by a certain output functional (solution).

We shall act analogously to the structure method for a linear inverse problem to determine the right side of the differential equation (the structure block input) by certain information on its exit.

Consider the following scheme. Let the inverse problem be the successive solving of two problems. The first problem is either the direct or inverse problem of determining the right side of differential operator L

$$LQ = q \tag{2.1}$$

with initial boundary conditions

$$MQ = a. \tag{2.2}$$

The operator M usually is the trace-following operator by spatial-time variables.

The second problem for the known operator P is connected with the first one as follows

$$Pu = Q, \tag{2.3}$$

$$Ru = b, \tag{2.4}$$

where $Ru = b$ are the (1.18) data.

Let S be a certain known operator, and S^{-1} exist. Solving (1.16)-(1.17) and (1.18)-(1.19) is equivalent to determining the functions u and F from the equations

$$L_1 u = F, \tag{2.5}$$

$$SF = q \tag{2.6}$$

by the data

$$Ru = b, \tag{2.7}$$

$$MPu = a, \tag{2.8}$$

where $L_1 = S^{-1}LP$.

Really, applying to (1.20) the operator S and using the fact that $SS^{-1} = I$ (the unit operator), we divide problem (1.20)-(1.23) for $Pu = Q$ into two successive problems (1.16)-(1.17) and (1.18)-(1.19). Conversely, applying L to (1.18) and substituting the right side as $q = SF$, we obtain the equivalent problem (1.20)-(1.23).

Thus, the block representation of solutions to direct and inverse problems allows us not only to construct new settings to determine the right side of the equations but to reduce the complicated problem to simpler ones.

Determining the harmonic oscillator. Let $S = P$ and suppose q is known. Such a scheme in the structure method is known as successive connecting of the structure blocks (Butkovskii, 1979). It was used for solving inverse problems by Bukhgeim (1983). We shall consider the inverse problem of determining the harmonic oscillator. This means that $q = 0$ and

$$P = \frac{d^2}{dt^2} + \omega^2,$$

where $\omega \neq 0$. Thus, the source is a harmonic oscillator of the form

$$F(x, z, t) = \bar{A}(x, z) \cos \omega t + \bar{B}(x, z) \sin \omega t$$

with known frequency ω. Here $x \in R^2$, $z \in R^1$, $t \in R^1_+$, and the bar denotes a three-dimensional vector.

Let L be the Lamé operator (Petrashen', 1978) which we shall denote as follows

$$LQ = Q_{zz} - \bar{A}Q_{tt} + L_0 Q = 0, \tag{2.9}$$

where L_0 is the linear differential operator of the second order by x and the first order by z. The Lamé constants $\lambda(x,z), \mu(x,z)$ and the medium density $\rho(x,z)$ entering in $\bar{A}$ are assumed to be analytic, known, and $\mu \neq 0, \rho \neq 0$. Coefficient $\bar{A}$ is as follows

$$\bar{A} = \begin{pmatrix} 1/v_s^2 & 0 & 0 \\ 0 & 1/v_s^3 & 0 \\ 0 & 0 & 1/v_p^2 \end{pmatrix},$$

where $v_s^2 = \mu/\rho$, $v_p^2 = (\lambda + 2\mu)/\rho$.

We must find data of the form (1.17), (1.19) for the operators P and L using the inverse problem data. Let the Cauchy data for equations (1.24) be as follows

$$\bar{Q}\Big|_{z=0} = \bar{v}_0(x,t), \tag{2.10}$$

$$\bar{Q}_z\Big|_{z=0} = \bar{v}_1(x,t). \tag{2.11}$$

For the equation

$$\frac{d^2\bar{u}}{dt^2} + \omega^2\bar{u} = \bar{Q}(x,z,t) \tag{2.12}$$

the data (1.19) may be analogous to (1.25), (1.26)

$$\bar{u}\Big|_{t=0} = \bar{\varphi}(x,z), \tag{2.13}$$

$$\bar{u}_t\Big|_{t=0} = \bar{\psi}(x,z). \tag{2.14}$$

As the operators L and P are commutative, we easily obtain $L_1 = L$ and, thus, the setting equivalent to problems (1.24)-(1.26) and (1.27)-(1.29) has the form

$$L\bar{u} = \bar{F}(x, z, t); \tag{2.15}$$

$$\bar{u}\Big|_{t=0} = \bar{\varphi}(x, z); \tag{2.16}$$

$$\bar{u}_t\Big|_{t=0} = \bar{\psi}(x, z); \tag{2.17}$$

$$PF = 0; \tag{2.18}$$

$$\left(\frac{d^2\bar{u}}{dt^2} + \omega^2\bar{u}\right)\Bigg|_{z=0} = \bar{v}_0(x, t); \tag{2.19}$$

$$\left(\frac{d^2\bar{u}}{dt^2} + \omega^2\bar{u}\right)_z\Bigg|_{z=0} = \bar{v}_1(x, t). \tag{2.20}$$

It is clear that if the data

$$\bar{u}\Big|_{z=0} = \bar{u}_0(x, t), \tag{2.21}$$

$$\bar{u}_z\Big|_{z=0} = \bar{u}_1(x, t) \tag{2.22}$$

are given, then $\bar{v}_0$ and $\bar{v}_1$ are calculated.

Let the following equalities hold:

$$\bar{u}_{tt} = w, \quad \bar{w}(x, 0, 0) = \bar{w}_t(x, 0, 0) = \bar{w}_z(x, 0, 0) = \bar{w}_{zt}(x, 0, 0) = 0.$$

Then the following theorem holds true.

Theorem 1.3. Let the functions $\bar{v}_0(x,t), \bar{v}_1(x,t), \bar{\varphi}(x,z), \bar{\psi}(x,z)$ be analytic in their variables and the concordance conditions hold $\omega^2\bar{\varphi}(x,0) = \bar{v}_0(x,0)$, $\omega^2\bar{\varphi}_z(x,0) = \bar{v}_1(x,0)$, $\omega^2\psi(x,0) = \bar{v}_{0t}(x,0)$, $\omega^2\bar{\psi}_z(x,0) = \bar{v}_{1t}(x,0)$. Then in the class of analytic functions, the inverse problem (1.31)-(1.35) has a unique solution

$$F(x,z,t) = \left(\bar{\varphi}_{zz} + L_0\bar{\varphi} + A\omega^2\bar{\varphi} - A\bar{Q}(x,z,0)\right)\cos\omega t$$

$$+ \left(\bar{\psi}_{zz} + L_0\bar{\psi} + A\omega^2\bar{\psi} - A\bar{Q}(x,z,0)\right)\sin\omega t, \tag{2.23}$$

where $Q(x,z,t)$ solves the Cauchy problem (1.24)-(1.26).

Proof. The proof of uniqueness and existence of solution to (1.24)-(1.26) is based on the Cauchy-Kovalevskaya theorem. As is known, the solution to (1.27) with data (1.28)-(1.29) has the form

$$\bar{u}(x,z,t) = \bar{\varphi}(x,z)\cos\omega t + \frac{1}{\omega}\bar{\psi}(x,z)\sin\omega t$$

$$+\frac{1}{\omega}\int_0^t \bar{Q}(x,z,\tau)\sin\omega(t-\tau)d\tau.$$

Substituting the expression in (1.30), taking into account (1.24) and twice integrating term-by-term, we obtain the desired representation (1.38).

This setting obviously may be generalized. So, instead of (1.32)-(1.33) we may consider more general conditions

$$\left.\frac{\partial^k \bar{u}}{\partial t^k}\right|_{t=t_0} = \bar{\varphi}(x,z), \tag{2.24}$$

$$\left.\frac{\partial^m \bar{u}}{\partial t^m}\right|_{t=t_1} = \bar{\psi}(x,z), \tag{2.25}$$

where $k=0,\ k=1,\ m=0,\ m=1$.

Let, further, $t \in R^1$ and the boundary conditions have the form (1.36)-(1.37). Then for inverse problem (1.30)-(1.31), (1.36)-(1.37), (1.39)-(1.40) the following theorem holds.

Theorem 1.4. Let the functions $\bar{u}_0(x,t), \bar{u}_1(x,t), \bar{\varphi}(x,z), \bar{\psi}(x,z)$ be analytic in their variables and the concordance conditions hold

$$\frac{\partial^k \bar{u}_1}{\partial t^k}\Big|_{t=t_0} = \bar{\varphi}(x,0); \quad \frac{\partial^k \bar{u}_1}{\partial t^k}\Big|_{t=t_0} = \bar{\varphi}_z(x,0);$$

$$\frac{\partial^m \bar{u}_0}{\partial t^m}\Big|_{t=t_1} = \bar{\psi}(x,0); \quad \frac{\partial^m \bar{u}_1}{\partial t^m}\Big|_{t=t_1} = \bar{\psi}_m(x,0).$$

Assume, also, that

$$t_0 - t_1 \neq \frac{2\pi l}{\omega}, \quad k = m, \tag{2.26}$$

$$t_0 - t_1 \neq \frac{\pi}{\omega}\left(l + \frac{1}{2}\right), \tag{2.27}$$

where $l = 0, \pm 1, \pm 2, \ldots$. Then in the class of analytic functions increasing rapidly at infinity the inverse problem has a unique solution

$$\bar{F}(x,z,t) = \left(A\omega^2 \bar{d}_1 + \bar{d}_{1zz} + L_0 \bar{d}_1\right)\cos\omega t$$

$$+\left(A\omega^2 \bar{d}_2 + \bar{d}_{2zz} + L_0 \bar{d}_2\right)\sin\omega t + \frac{A}{\omega}\int\limits_{-\infty}^{t} \bar{Q}(x,z,\tau)\sin\omega(t-\tau)d\tau,$$

where $Q(x,z,t)$ solves the Cauchy problem (1.24), (1.36)-(1.37), and the functions $\bar{d}_1(x,z), \bar{d}_2(x,z)$ are as follows

$$\bar{d}_1(x,z) = \left[b_k(t_0)\bar{\psi}(x,z) - b_m(t_1)\bar{\varphi}(x,z) + \frac{1}{\omega}\left(b_k(t_0)\int\limits_{-\infty}^{t_1} \bar{Q}(x,z,\tau)\right.\right.$$

$$\left.\left.\times c_m(t_1-\tau)d\tau - b_m(t_1)\int\limits_{-\infty}^{t_0} \bar{Q}(x,z,\tau)c_k(t_0-\tau)d\tau\right)\right]$$

$$\times\left[a_k(t_0)b_m(t_1) - b_k(t_0)a_m(t_1)\right]^{-1};$$

$$\bar{d}_2(x,z) = -\left[a_k(t_0)\bar{\psi}(x,z) - a_m(t_1)\bar{\varphi}(x,z) + \frac{1}{\omega}\left(a_k(t_0)\int\limits_{-\infty}^{t_1} \bar{Q}(x,z,\tau)\right.\right.$$

$$\times c_m(t_1-\tau)d\tau + a_m(t_1)\int\limits_{-\infty}^{t_0} \bar{Q}(x,z,\tau)c_k(t_0-\tau)d\tau\Bigg)\Bigg]$$

$$\times\left[a_k(t_0)b_m(t_1)-b_k(t_0)a_m(t_1)\right]^{-1};$$

$$a_j(t)=(1-j)\cos\omega t - j\omega^j\sin\omega t;$$

$$b_j(t)=(1-j)\sin\omega t + j\omega^j\cos\omega t;$$

$$c_j(t-\tau)=(1-j)\sin\omega(t-\tau)+j\omega^j\cos\omega t;\quad j=0,1.$$

Proof. The proof is analogous to the proof of theorem 1.3. Conditions (1.41)-(1.42) denote that the denominators in expressions for $\bar{d}_1$ and $\bar{d}_2$ are not equal to zero.

Formulas for problems of source determination. Let L and M be such differential operators that

$$LM = ML. \tag{2.28}$$

Consider the equation

$$Lu = f, \tag{2.29}$$

where f is such that

$$Mf = 0. \tag{2.30}$$

For the function u depending on the form of L and M we have initial and boundary conditions. The problem is to find the source f given the data for u.

Formulas for determining f for concrete L and M are given in Ayupova (1986). The algorithm is based on the structure method of inverse problem solving (Erokhin, 1984). Apply to both sides of (1.44) the operator M. By (1.43) and (1.45) we obtain

$$LMu = 0.$$

Denoting $Mu = w$ we obtain

$$Lw = 0.$$

Successively solving equations $Lw = 0$ and $Mu = w$, we obtain u. Substituting u into (1.44), we obtain f. Denote, for convenience

$$\Delta = \sum_{i=1}^{n} \frac{\partial^2}{\partial x_i^2}, \quad \Delta_{n-i} = \sum_{\substack{j=1 \\ j \neq i}}^{n} \frac{\partial^2}{\partial x_j^2},$$

$$\bar{\xi} = \Big(\xi_1, \dots, \xi_{i-1}, \xi_{i+1}, \dots, \xi_n\Big), \quad \tilde{x} = \Big(x_1, \dots, x_{i-1}, x_{i+1}, \dots, x_n\Big).$$

Let us consider some examples (Ayupova, 1986).

Example 1.

$$Lu = u_t - \Delta u = f(x,t),$$

$$Mf = f_{x_i} - \Delta_{n-1} f = 0, \quad i \in R^+, \quad x \in R^n.$$

$$u\Big|_{t=0} = u_0(x), \qquad u\Big|_{x_i=0} = p_0(\tilde{x}, t),$$

$$f(x,t) = \frac{1}{(2\sqrt{\pi x_i})^{n-1}} \int\limits_{R^{n-1}} \Big[p_{0t} - p_{0\tilde{\xi}\tilde{\xi}} - p_{0\tilde{\xi}}^{(4)} - w_{\xi_i}(0, \tilde{\xi}, t)\Big]$$

$$\times \exp\left(-\frac{|\tilde{x} - \tilde{\xi}|^2}{4x_i}\right) d\tilde{\xi} - - \int\limits_{R^{n-1}} w(0, \tilde{\xi}, t) \exp\left(-\frac{|\tilde{x} - \tilde{\xi}|^2}{4x_i}\right)$$

$$\times \left[|\tilde{x} - \tilde{\xi}|^2 \left(\sqrt{x_i}\right)^{-(n+3)} - \frac{n-1}{2} \left(\sqrt{x_i}\right)^{-(n+1)}\right] d\tilde{\xi},$$

where $x_i \in R^+$ and w solves the problem

$$w_t - \Delta w = 0,$$

$$w\Big|_{t=0} = u_{0x_i} - \Delta_{n-1} u_0.$$

Example 2.

$$Lu = u_t - \Delta u = f(x,t),$$

$$Mf = f_{x_i x_i} + \omega^2 f = 0, \quad t \in R^+, \quad x \in R^n.$$

$$u\Big|_{t=0} = u_0(x), \quad u\Big|_{x_i=0} = p_0(\tilde{x},t), \quad u_{x_i}\Big|_{x_i=0} = p_1(\tilde{x},t),$$

$$f(x,t) = a(t,\tilde{x})\cos\omega x_i + b(t,\tilde{x})\sin\omega x_i,$$

where

$$a(t,\tilde{x}) = -w(0,\tilde{x},t) + p_{0t}(t,\tilde{x}) - \sum_{\substack{j=1\\ j\neq i}}^{n} \frac{\partial^2}{\partial x_j^2} p_0(t,\tilde{x}) + \omega p_0(t,\tilde{x}),$$

$$b(t,\tilde{x}) = -\frac{1}{\omega} w_{x_i}(0,\tilde{x},t) + \frac{p_{1t}(\tilde{x},t)}{\omega} - \frac{1}{\omega}\sum_{\substack{j=1\\ j\neq i}}^{n} \frac{\partial^2}{\partial x_j^2} p_1(t,\tilde{x}) + \omega p_1(t,\tilde{x}),$$

solves the problem

$$w_t - \Delta w = 0, \qquad w\Big|_{t=0} = u_{0x_i x_i} + \omega^2 u_0.$$

Example 3.

$$Lu = u_t - \Delta u = f(x,t),$$

$$Mf = f_{x_i x_i} - f_{x_j x_j} = 0, \quad t \in R^+, \quad x \in R^n.$$

$$u\Big|_{t=0} = u_0(x), \quad u\Big|_{x_i=0} = p_0(\tilde{x},t), \quad u_{x_i}\Big|_{x_i=0} = p_1(\tilde{x},t),$$

$$f(x,t) = \frac{1}{2}\int_0^{x_i} \Big[w_{x_i}\Big(\xi_i, x_j + (x_i - \xi_i), x', t\Big) - w_{x_j}\Big(\xi_i, x_j + (x_i - \xi_i), x', t\Big)\Big] d\xi_i$$

$$+\frac{1}{2}\int_{x_j - x_i}^{x_j + x_i} \sum_{\substack{k=1\\ k\neq i,j}}^{n} \frac{\partial^2}{\partial x_k^2} p_1(\xi_j, x', t)\, d\xi_i$$

$$-\frac{1}{2}\left[\sum_{\substack{k=1\\ k\neq i,j}}^{n} \frac{\partial^2}{\partial x_k^2} p_0\Big(x_j + x_i, x', t\Big) + \sum_{\substack{k=1\\ k\neq i,j}}^{n} \frac{\partial^2}{\partial x_k^2} p_0(x_j - x_i, x', t)\right]$$

$$- \left[p_{1x_j}\left(x_j + x_i, x', t\right) - p_{1x_j}\left(x_j - x_i, x', t\right)\right]$$

$$+ \left[p_{0x_jx_j}\left(x_j + x_i, x', t\right) - p_{0x_jx_j}\left(x_j - x_i, x', t\right)\right],$$

where $x' = (x_1, \ldots, x_{i-1}, x_{i+1}, \ldots, x_{j-1}, x_{j+1}, \ldots, x_n)$ and w solves the problem

$$w_t - \Delta w = 0,$$

$$w\Big|_{t=0} = u_{0x_ix_i} - u_{0x_jx_j}.$$

Example 4.

$$Lu = u_t - \Delta u = f(x, t),$$

$$Mf = \Delta f = 0, \quad t \in R^+, \quad x \in R^n.$$

$$u\Big|_{t=0} = u_0(x), \quad u\Big|_{\partial S} = p_0(x, t).$$

For $t = 0$ the concordance condition $u_0 = p_0(x, 0)$ must hold. S is the sphere of the radius R in the space R^n.

$$f(x, t) = \int\limits_{|\xi|=R} p_R(x, \xi) p_{0t}(\xi, t) dS_\xi$$

$$- \int\limits_{|\xi|<R} G_R(x, \xi) \left[-\frac{n}{(\sqrt{\pi})^n (2\sqrt{t})^{n+1}} \int\limits_{R^n} \Delta u_0(\eta) \exp\left(-\frac{|\xi - \eta|^2}{4t}\right) d\eta \right.$$

$$\left. + \frac{1}{(2\sqrt{\pi t})^n) 4t^2} \int\limits_{R^n} |\xi - \eta|^2 \Delta u_0(\eta) \exp\left(-\frac{|\xi - \eta|^2}{4t}\right) d\eta \right] d\xi$$

$$- \frac{1}{(2\sqrt{\pi t})^n)} \int\limits_{R^n} \Delta u_0(\xi) \exp\left(-\frac{|x - \xi|^2}{4t}\right) d\xi$$

for $x \in R$, where

$$p_R(x, \xi) = \frac{R^2 - |x|^2}{\sigma_n R |x - \xi|^n};$$

$$G_R(x,\xi) = \frac{1}{\sigma_n}\left(\frac{1}{|x-\xi|^{n-2}} - \frac{(R/|x|)^{n-2}}{|(R/|x|)^2 x - \xi|^{n-2}}\right);$$

σ_n is the area of the unit sphere in R^n.

Example 5.

$$Lu = u_{tt} - \Delta u = f(x,t),$$

$$Mf = f_{x_i} - \Delta_{n-1} f = 0, \quad t \in R^+, \quad x \in R^n.$$

$$u\Big|_{t=0} = u_0(x), \quad u\Big|_{x_i=0} = p_0(\tilde{x},t), \quad u_t\Big|_{t=0} = u_1(x),$$

$$f(x,t) = \int_0^{x_i}\int_{R^{n-1}}\left\{\frac{w_{\xi_i\xi_i}(\xi_i,\tilde{\xi},t)}{\left[2\sqrt{\pi(x_i-\xi_i)}\right]^{n-1}}\right.$$

$$\left.+\frac{\Delta_{n-1}\Delta_{n-1}w(\xi_i,\tilde{\xi},t)}{\left[2\sqrt{\pi(x_i-\xi_i)}\right]^{n-1}}\right\}\exp\left(-\frac{|\tilde{x}-\tilde{\xi}|^2}{4x_i}\right)d\tilde{\xi}d\xi$$

$$+\frac{1}{(2\sqrt{\pi x_i})^{n-1}}\int_{R^{n-1}}\left[\varphi_{0tt} - \Delta_{n-1}\varphi_0(\tilde{\xi},t)\right.$$

$$\left.+\Delta_{n-1}\Delta_{n-1}p_0(\tilde{\xi},t)\right]\exp\left(-\frac{|\tilde{x}-\tilde{\xi}|^2}{4x_i}\right)d\tilde{\xi} - w_{x_i} - \Delta_{n-1}w$$

For $x_i \in R^+$, where w solves the problem

$$w_{tt} - \Delta w = 0,$$

$$w\Big|_{t=0} = u_{0x_i} - \Delta_{n-1}u_0,$$

$$w_t\Big|_{t=0} = u_{1x_i} - \Delta_{n-1}u_1.$$

Example 6.

$$Lu = u_{tt} - \Delta u = f(x,t),$$

$$Mf = f_{x_i x_i} + \omega^2 f = 0, \quad t \in R^+, \quad x \in R^n.$$

$$u\Big|_{t=0} = u_0(x), \quad u\Big|_{x_i=0} = p_0(\tilde{x},t),$$

$$u_t\Big|_{t=0} = u_1(x), \quad u_{x_i}\Big|_{x_i=0} = p_1(\tilde{x},t),$$

$$f(x,t) = a(t,\tilde{x})\cos\omega x_i + b(t,\tilde{x})\sin\omega x_i,$$

where

$$a(t,\tilde{x}) = p_{0tt} - p_{0\tilde{x}\tilde{x}} + p_0\omega^2 - w(0,\tilde{x},t),$$

$$b(t,\tilde{x}) = \frac{p_{1tt}}{\omega} - \frac{p_{1\tilde{x}\tilde{x}}}{\omega} + \omega p_1 + \frac{1}{\omega} w_{x_i}(0,\tilde{x},t),$$

and w solves the problem

$$w_{tt} - \Delta w = 0,$$

$$w\Big|_{t=0} = u_{0x_i x_i} + \omega^2 u_0,$$

$$w_t\Big|_{t=0} = u_{1x_i x_i} + \omega^2 u_1.$$

Example 7.

$$Lu = u_{tt} - \Delta u = f(x,t),$$

$$Mf = f_{x_i x_i} - f_{x_j x_j} = 0, \quad t \in R^+, \quad x \in R^n.$$

$$u\Big|_{t=0} = u_0(x), \quad u\Big|_{x_i=0} = p_0(\tilde{x},t),$$

$$u_t\Big|_{t=0} = u_1(x), \quad u_{x_i}\Big|_{x_i=0} = p_1(\tilde{x},t).$$

$$f(x,t) = \frac{1}{2}\int_0^{x_i} \Big[w_{x_j}\Big(\xi_i, x_j + (x_i - \xi_i), x', t\Big)$$

$$-w_{x_j}\Big(\xi_i, x_j-(x_i-\xi_i), x', t\Big)\Big]d\xi_i + \frac{1}{2}\int\limits_{x_j-x_i}^{x_j+x_i} p_{1tt}(\xi_j, x', t)d\xi_j$$

$$+\frac{1}{2}\Big[p_{0tt}(x_0+x_i, x', t) + p_{0tt}(x_0-x_i, x', t)\Big]$$

$$-\frac{1}{2}\int\limits_{x_j-x_i}^{x_j+x_i} \sum_{\substack{k=1\\ k\neq i,j}}^{n} \frac{\partial^2}{\partial x_k^2} p_1(\xi_j, x', t)d\xi_j$$

$$-\frac{1}{2}\left[\sum_{\substack{k=1\\ k\neq i,j}}^{n}\left(\frac{\partial^2}{\partial x_k^2}p_0(x_j+x_i, x', t) + \frac{\partial^2}{\partial x_k^2}p_0(x_j-x_i, x', t)\right)\right]$$

$$-\Big[p_{1x_j}(x_j+x_i, x', t) - p_{1x_j}(x_j-x_i, x', t)\Big]$$

$$-\Big[p_{0x_jx_j}(x_j+x_i, x', t) + p_{0x_jx_j}(x_j-x_i, x', t)\Big],$$

where w solves the problem

$$w_{tt} - \Delta w = 0,$$

$$w\Big|_{t=0} = u_{0x_ix_i} - u_{0x_jx_j},$$

$$w_t\Big|_{t=0} = u_{1x_ix_i} - u_{1x_jx_j}.$$

Example 8.

$$Lu = u_{tt} - \Delta u = f(x,t),$$

$$Mf = \Delta f = 0, \quad t \in R^+, \quad x \in R^n.$$

$$u\Big|_{t=0} = u_0(x), \quad u\Big|_{\partial S} = p_0(x,t), \quad u_t\Big|_{t=0} = u_1(x),$$

S is the sphere of radius R in the space R^n.

$$f(x,t) = \int\limits_{|\xi|=R} p_R(x,\xi) p_{0tt}(\xi,t)d\xi - \int\limits_{|\xi|<R} G_R(x,\xi) w_{tt}(\xi,t)d\xi - w(x,t).$$

For $x \in S$, $G_R(x,\xi), p_R(x,\xi)$ from example 4, w solves the problem

$$w_{tt} - \Delta w = 0,$$

$$w_t\Big|_{t=0} = \Delta u_1(x),$$

$$w\Big|_{t=0} = \Delta u_0(x).$$

Example 9.

$$Lu = u_t - u_{xx} = f(x,t),$$

$$Mf = f_{xx} + \omega^2 f = 0, \quad t \in R^+, \quad x \in R.$$

$$u\Big|_{t=0} = u_0(x), \quad u\Big|_{x=0} = p_0(t), \quad u_x\Big|_{x=0} = p_1(t),$$

$$f(x,t) = a(t)\cos\omega x + b(t)\sin\omega x,$$

where

$$a(t) = p_0(t) + p_0(t)\omega^2 - w(0,t),$$

$$b(t) = \frac{p_1}{\omega} + \omega p_1(t) - \frac{1}{\omega} w_x(0,t),$$

w solves the problem

$$w_t - w_{xx} = 0,$$

$$w\Big|_{t=0} = u_{0xx} + \omega^2 u_0.$$

Example 10.

$$Lu = u_{tt} - u_{xx} = f(x,t),$$

$$Mf = f_{xx} + \omega^2 f = 0, \quad t \in R^+, \quad x \in R.$$

$$u\Big|_{t=0} = u_0(x), \quad u\Big|_{x=0} = p_0(t),$$

$$u_t\Big|_{t=0} = u_1(x), \quad u_x\Big|_{x=0} = p_1(x),$$

$$f(x,t) = a(t)\cos\omega x + b(t)\sin\omega x,$$

where

$$a(t) = -w(0,t) + \omega^2 p_0(t) + p_0''(t),$$

$$b(t) = -\frac{1}{\omega} w_x'(0,t) + \omega p_1(t) + \frac{p_1''(t)}{\omega},$$

w solves the problem

$$w_{tt} - w_{xx} = 0,$$

$$w\Big|_{t=0} = u_{0xx} + \omega^2 u_0,$$

$$w_t\Big|_{t=0} = u_{1xx} + \omega^2 u_1.$$

§1.3 Determining the index and the right side of the pseudodifferential equation.

Let R^{n+2} be the $(n+2)$-dimensional real Euclidean space of the variables $x, y, t, x \in R^n$, $y \in R^1$, $t \in R^1$. Consider the equation

$$\frac{\partial^2 u}{\partial y^2} + Lu = g(y)f(x,t), \tag{3.1}$$

where L is the integral operator

$$Lu = \int\limits_{R^{n+1}} K(x-\xi, t-\tau)u(\xi,\tau)d\xi d\tau. \tag{3.2}$$

Such operators L are pseudodifferential and arise from investigation of waves in linear dispersing media (Trev, 1984). In particular, for

$$K(x,t) = \sum_{\alpha,\beta} c_{\alpha\beta}\delta^{(\alpha)}(x)\delta^{(\beta)}(t),$$

the operator L is equivalent to the differential operator with constant coefficients

$$L = \sum_{\alpha,\beta}(-1)^{|\alpha|+\beta} c_{\alpha\beta} D_x^\alpha D_t^\beta.$$

The problem is to find $u(x,y,t)$, $f(x,t)$ and the operator L of the form (1.47) if the functions $g(y)$ are known,

$$u_0(x,t) = u(x,y,t)\Big|_{y=0}, \tag{3.3}$$

$$u_1(x,t) = u_y(x,y,t)\Big|_{y=0}. \tag{3.4}$$

Note, that parallel with the function $f(x,t)$, characterizing the source, the operator L is also unknown. It is the novelty of the given setting (Anikonov and Erokhin, 1984). This is so because, in general wave theory, the equations describing the wave motion are often unknown. So, Whizem (1977) notes that unlike for nondispersing waves, for the waves in linear dispersing media, the problem of which equation they satisfy and how to solve it is very complicated.

Let $S(R^n)$ be the space of rapidly decreasing functions; $S'(R^n)$ be the space of generalized functions of slow growth; let $\mathcal{E}'(R^n)$ be the space of generalized functions with compact support; let Θ_M be the set of multiplicators in S, i.e., the set of infinitely differentiable functions in R^n growing at infinity not faster than a polynomial (Vladimirov, 1976).

With the special representation of the desired operator in mind, it is convenient to formulate the above setting of the inverse problem in terms of the Fourier images of functions.

Let the Fourier image of $\varphi \in S(R^{n+1})$ be as follows

$$F[\varphi] = \hat{\varphi}(k,\omega) = \int\limits_{R^{n+1}} \varphi(x,t) e^{i(\langle k,x\rangle - \omega t)} dx dt,$$

where $\langle k, x\rangle$ is scalar product of the vectors $k = (k_1, \dots, k_n)$ and $x = (x_1, \dots, x_n)$.

Assume that the solution to (1.46), the right side of (1.46), the function $K(x,t)$, and data (1.48)-(1.49) undergo Fourier transformation by x and t. Let $u' = u'(y,k,\omega), \hat{f}, \hat{u}_0, \hat{u}_1, \hat{K}$ be the Fourier transforms of u, f, u_0, u_1, K respectively. Then problem (1.48)-(1.49) becomes as follows. To find the functions $u'(y,k,\omega)$, $\hat{K}(k,\omega)$, and $\hat{f}(k,\omega)$ if

$$\frac{d^2u'}{dy^2} + (-1)^{n+1}\hat{K}(k,\omega)u' = g(y)\hat{f}(k,\omega), \tag{3.5}$$

$$u'\Big|_{y=0} = \hat{u}_0(k,\omega), \tag{3.6}$$

$$u'_y\Big|_{y=0} = \hat{u}_0(k,\omega). \tag{3.7}$$

Further, we suppose that $\mathrm{Re}(\hat{K}^{1/2}) \neq 0$. This condition holds for the wave equation with absorption (Erokhin, 1982).

Theorem 1.5. Let $u_0(x,t), u_1(x,t) \in S(R^{n+1})$; and let the solution $u(x,y,t)$ to (1.46) belong to $S'(R^{n+2}), g(y) \in \mathcal{E}'(R^1)$, where

$$\{\mathrm{supp}\, g(y)\} \bigcap \{y = 0\} = \{\emptyset\}.$$

Assume, also, that $\hat{K}(k,\omega)$ and the ratio $\hat{u}_1(k,\omega)/\hat{u}_0(k,\omega)$ belong to Θ_M. If $\mathrm{Re}(\hat{u}_1)\mathrm{Re}(u_0) + \mathrm{Im}(\hat{u}_1)\mathrm{Im}(u_0) \neq 0$ and $g(y) + g(-y) \neq 0$, then inverse problem (1.46)-(1.49) has a unique solution determined in the spectral domain by the formulas

$$\hat{K}(k,\omega) = (-1)^n \frac{\hat{u}_1^2(k,\omega)}{4\hat{u}_0^2(k,\omega)},$$

$$\hat{f}(k,\omega) = \frac{\hat{u}_1}{\int\limits_{R^1} g(\eta)\exp\left(\pm|\eta|\sqrt{(-1)^n\hat{K}}\right) d\eta},$$

$$u'(y,k,\omega) = \pm\frac{\hat{f}}{2\sqrt{(-1)^n\hat{K}}}\int_{R^1} g(\eta)e^{\pm|y-\eta|\sqrt{(-1)^n\hat{K}}}\,d\eta,$$

where $(+)$ corresponds to the case $\mathrm{Re}\left(\sqrt{(-1)^n\hat{K}}\right) < 0$ and $(-)$ corresponds to the case $\mathrm{Re}\left(\sqrt{(-1)^n\hat{K}}\right) > 0$.

Proof. The general solution to homogeneous equation (1.50) is as follows:

$$v(y,k,\omega) = c_1 e^{-yW} + c_2 e^{yW},$$

where c_1 and c_2 are arbitrary functions depending on k and ω; $W(k,\omega) = \sqrt{(-1)^n\hat{K}(k,\omega)}$. By assumption, $\mathrm{Re}(W) \neq 0$. Therefore, the fundamental solution is

$$G(y,k,\omega) = \pm\frac{1}{2}\frac{e^{\pm|y|W}}{W}.$$

Here $(+)$ corresponds to the case $\mathrm{Re}(W) < 0$ and $(-)$ corresponds to the case $\mathrm{Re}(W) > 0$. Thus, the general solution of (1.50) is as follows:

$$u'(y,k,\omega) = c_1 e^{-yW} + c_2 e^{yW} \pm \frac{\hat{f}}{2W}\int_{R^1} g(\eta)e^{\pm|y-\eta|W}\,d\eta. \tag{3.8}$$

The convolution in (1.53) exists, since, by assumption $g(y)$ has a compact support. As $u'(y,k,\omega) \in S'(R^n)$ and the first two terms exponentially grow as $|y| \to \infty$, then $c_1 = c_2 = 0$. Since the function support is separated from zero, data (1.59) and (1.52) are determined by relations

$$\hat{u}_0(k,\omega) = \pm\frac{\hat{f}}{2W}\int_{R^1} g(\eta)e^{\pm|\eta|W}\,d\eta,$$

$$\hat{u}_1(k,\omega) = \hat{f}\int_{R^1} g(\eta)e^{\pm|\eta|W}\,d\eta. \tag{3.9}$$

This implies that $\hat{u}_0(k,\omega) = \pm\frac{\hat{u}_1(k,\omega)}{2W(k,\omega)}$.

By assumption, $W(k,\omega), \hat{u}_1/\hat{u}_0 \in \Theta_M$ and $\hat{u}_0, \hat{u}_1 \in S(R^{n+1})$; therefore, from the latter equality we correctly determine the function $\hat{K}(k,\omega)$

$$\hat{K}(k,\omega) = (-1)^n \frac{\hat{u}_1^2(k,\omega)}{4\hat{u}_0^2(k,\omega)}.$$

Substituting the obtained function $W = \sqrt{(-1)^n \hat{K}}$ into (1.54) and taking account of the fact that $g(y) + g(-y) \neq 0$ and $\mathrm{Re}(W) \neq 0$, we uniquely obtain $\hat{f}(k,\omega)$

$$\hat{f}(k,\omega) = \frac{\hat{u}_1}{\int\limits_{R^1} g(\eta)\exp\left(\pm|\eta|\sqrt{(-1)^n\hat{K}}\right)d\eta}.$$

Thus, after determining W and $\hat{f}$, the general solution to (1.46) is determined by (1.53). The theorem is proved.

Remark. Analogous results may be obtained for inverse problems for other representations of right sides of equation (1.46). For example, in Erokhin (1982), for the wave equation were considered

$$\mathcal{F}_k(x,y,t) = g_k(x_k) - c_0 y - \sum_{\substack{j=1\\k\neq j}}^{n}(c_j x_j - c_{n+1}t) f_k(x',y,t), \tag{3.10}$$

$$\mathcal{F}_{n+1}(x,y,t) = g_{n+1}\left(t - c_0 y - \sum_{j=1}^{n} c_j x_j\right) f_{n+1}(x,y), \tag{3.11}$$

where $k = 1,\dots,n$; $x' = x_1,\dots,x_{k-1},x_{k+1},\dots,x_n$. The constants $c_k, k = 0,\dots,n+1$ and the functions are assumed to be known.

Remark. The stated method may be applied also for the case when (1.46) is considered either in the half-space $y > 0$, or in the layer $0 < y < l$. So, for the half-space the setting analogous to (1.46)-(1.49) may be executed as follows. Equation (1.46) is considered for $y > 0$, and its right side may also be described by functions (1.55)-(1.56). It is necessary to find the function f_k and the operator L of the form (1.47) using the data

$$u\Big|_{y=0} = u_0(x,t), \quad u_y\Big|_{y=0} = u_1(x,t), \quad u\Big|_{y=h} = u_h(x,t).$$

§1.4 Symmetry in inverse problems of source determination.

Investigation of inverse problems connected with determining the source characteristics of various physical fields leads to group analysis methods for initial information. Under some conditions, a certain "symmetry" in the measured information corresponds to a certain "symmetry" of the desired source. Determining the properties of sources from the properties of the given information has obvious practical and theoretical uses. The method described in this section is based on two physical suppositions.

1. The source event is bounded and its domain does not intersect the domain of measurements.

2. The medium has a certain symmetry.

Now, let us give the precise formulations.

In a domain $D \subset R^n$, $n \geq 2$, $0 \leq t \leq t_0$ of variables $\bar{x}$, consider a linear boundary-value problem with homogeneous and initial conditions for the vector equation

$$\frac{\partial^2 \bar{u}}{\partial t^2} = L\bar{u} + \bar{F}(\bar{x},t), \tag{4.1}$$

where L is a linear differential operator of the second order.

Assume that the solution $\bar{u}(\bar{x},t)$ may be represented as follows

$$\bar{u}(\bar{x},t) = \int_0^t \int_D G(\bar{x},\bar{x}',t,\tau)\bar{F}(\bar{x}',\tau)d\bar{x}'d\tau. \tag{4.2}$$

Here $G(\bar{x},\bar{x}',t,\tau)$ is the Green function, $\bar{F}(\bar{x},t)$ is the source we need to determine.

Information is given by the vector functions

$$\bar{u}_0(\bar{s},t) = \bar{u}\Big|_{\Gamma}, \quad \bar{u}_1(\bar{s},t) = \frac{\partial \bar{u}}{\partial n}\Big|_{\Gamma}, \quad \bar{s} \in \Gamma, \tag{4.3}$$

where $\Gamma \subset D$ is a given hypersurface. Further, we must assume that all the functions are sufficiently smooth.

The basic theoretical assumptions that correspond to physical suppositions 1, 2 are formulated below.

1. The vector function $\bar{F}$ is finite in $\bar{x}$ for $t > 0$, is sufficiently differentiable and $suppF$ does not intersect with the surface Γ where the initial information (1.59) is given.

2. The Green function $G(\bar{x}, \bar{x}', \tau, t)$ satisfies the equation

$$M_{j\bar{x}}G = S_{j\bar{x}'}G, \quad j = 1, 2, \dots, m, \tag{4.4}$$

where $M_{j\bar{x}}$, $S_{j\bar{x}'}$ are certain linear differetial operators with constant coefficients acting by variables $\bar{x}$ and $\bar{x}'$ respectively.

Denote, further, by $M_{\bar{x}}^{\alpha}$ and $S_{\bar{x}'}^{\alpha}$ the set of operators of the form $\{M_{1\bar{x}}^{\alpha_1}, \dots, M_{m\bar{x}}^{\alpha_m}\}$ and $\{S_{1\bar{x}'}^{\alpha_1}, \dots, S_{m\bar{x}'}^{\alpha_m}\}$, $|\alpha| \leq N$ respectively. Here α is the multi-index. Evidently, the following statements hold.

Lemma 1. 1. Let $P_{\bar{x}} = \sum\limits_{|\alpha|\leq N} c_\alpha M_{\bar{x}}^{\alpha}$, $Q_{\bar{x}'} = \sum\limits_{|\alpha|\leq N} c_\alpha S_{\bar{x}'}^{\alpha}$, where c_α is a certain (scalar or vector) constant. Then

$$P_{\bar{x}} = Q_{\bar{x}'}G.$$

Lemma 1. 2. Let $P_{\bar{x}} = \sum\limits_{|\alpha|\leq N} c_\alpha M_{\bar{x}}^{\alpha}$, $Q_{\bar{x}'} = \sum\limits_{|\alpha|\leq N} c_\alpha S_{\bar{x}'}^{\alpha}$, where c_α is a certain (scalar or vector) constant. Then

$$P_{\bar{x}}\bar{u} = \int\limits_0^t \int\limits_D G(\bar{x}, \bar{x}', \tau, t) Q_{\bar{x}'}^t \bar{F}(\bar{x}', \tau) d\bar{x}' d\tau,$$

where $Q_{\bar{x}'}^t$ is the operator transposed to $Q_{\bar{x}'}$.

Lemma 1. 3. Let a surface Γ be non-characteristic; then the value of each differential operator $P_{\bar{x}}$ of the order N on the solution $\bar{u}(\bar{x}, t)$ for $\bar{x} \in \Gamma$ is uniquely determined by the values $\bar{u}_0$, $\bar{u}_1$ from (1.59).

Really, since, by the assumption, L is a second order operator, Γ is non-characteristic, and $supp\bar{F} \cap \Gamma = \{\emptyset\}$, then the normal to Γ derivative of solution $\bar{u}(\bar{x})$ is calculated from homogeneous equation (1.57) only *via* $\bar{u}_0(\bar{s}, t)$ and $u_1(\bar{s}, t)$, $\bar{s} \in \Gamma$. Therefore, $P_{\bar{x}}\bar{u}\Big|_{\Gamma}$ is uniquely determined by values of $\bar{u}_0$, $\bar{u}_1$ and the operator L.

Denote the contraction of $P_{\bar{x}}$ in Γ by $\tilde{P}_{\bar{x}}(\bar{u}_0, \bar{u}_1)$.

Lemma 1. 4. Let $P_{\bar{x}} = \sum\limits_{|\alpha| \leq N} c_\alpha M_{\bar{x}}^\alpha$, $Q_{\bar{x}'} = \sum\limits_{|\alpha| \leq N} c_\alpha S_{\bar{x}'}^\alpha$, where c_α is a certain (scalar or vector) constant. Then the following basic relation

$$\tilde{P}_{\bar{x}}(\bar{u}_0, \bar{u}_1) = \int\limits_0^t \int\limits_D G(\bar{s}, \bar{x}', \tau, t) Q_{\bar{x}}^t \bar{F}(\bar{x}', \tau) d\bar{x}' d\tau, \quad \bar{s} \in \Gamma$$

holds true.

This lemma follows from lemmas 1.2 and 1.3. For a wide class of G and $\bar{F}$ the following theorem holds.

Theorem 1.6. In order that $Q_{\bar{x}}^t \bar{F} = 0$ is necessary, and for certain restrictions imposed on G and $\bar{F}$, it is sufficient that

$$\tilde{P}_{\bar{x}}(\bar{u}_0, \bar{u}_1) = 0.$$

That is, the symmetry in the information $\tilde{P}_{\bar{x}}(\bar{u}_0, \bar{u}_1) = 0$ implies a certain symmetry of the function describing the field sources $Q_{\bar{x}}^t \bar{F}(\bar{x}) = 0$.

As an example, consider the system of equations of elasticity theory which describes the wave process in an ideal elastic infinite homogeneous space ($n = 3$). After Fourier transformation by the time variable we obtain a second order equation system of the form (1.60), where parameter ω denotes the frequency.

Let Γ be a sphere of radius R. Then, by the point 1, support of the source $\bar{F}(\bar{x}, \omega)$ must lie strictly inside it. Now, the Green function of this system depends on the argument difference $\bar{x}-\bar{x}' : G(\bar{x},\bar{x}',\omega) = G_0(\bar{x}-\bar{x}',\omega)$; therefore, we may take

$$M_{j\bar{x}} = \frac{\partial}{\partial x_j}, \quad S_{j\bar{x}'} = -\frac{\partial}{\partial x'_j}, \quad j=1,2,3,$$

which will satisfy the relation

$$\frac{\partial}{\partial x_j} G_0(\bar{x}-\bar{x}',\omega) = -\frac{\partial}{\partial x'_j} G_0(\bar{x}-\bar{x}',\omega), \quad j=1,2,3.$$

On the basis of such operators we construct different test operators $P_{\bar{x}}, Q_{\bar{x}'}$.

$$P_{\bar{x}} = \sum_{|\alpha|\le N} c_\alpha M^\alpha_{\bar{x}}, \qquad Q_{\bar{x}'} = \sum_{|\alpha|\le N} c_\alpha S^\alpha_{\bar{x}'}.$$

Depending on whether the c_α are scalar or vector, we may create different scalar (of div type), or vector (of rot type) differential operators over vector functions $\bar{u}_0, \bar{u}_1$, which are traces of the solution

$$\bar{u}_1(\bar{x},\omega) = \int\limits_{|x|\le R} \left\{ -\frac{e^{ik_1 r}}{r}\frac{1}{2\pi\rho\omega^2}\text{grad div}\bar{F}(\bar{x}',\omega) \right.$$

$$\left. + \frac{e^{ik_2 r}}{r}\left[\frac{1}{2\pi\mu}\bar{F}(\bar{x}',\omega) + \frac{1}{2\pi\rho\omega^2}\text{grad div}\bar{F}(\bar{x}',\omega)\right]\right\} d\bar{x}',$$

where $k_1^2 = \frac{\rho\omega^2}{\lambda+2\mu}$, $k_2^2 = \frac{\rho\omega^2}{\mu}$; ρ is the density; λ,μ are the Lamé parameters; $r^2 = (x_1-x'_1)^2 + (x_2-x'_2)^2 + (x_3-x'_3)^2$.

§1.5 Inversion formulas for the wave equation. Setting of the problem.

Consider the equation

$$\Delta u = \frac{1}{c^2}\frac{\partial^2 u}{\partial t^2} - \beta\frac{\partial u}{\partial t} - \alpha^2 u = f(x,y,t)g(z), \tag{5.1}$$

where $\alpha \neq 0$, $\beta \neq 0$, $x, y, z, t \in R^1$; the function $f(x, y, t)$ is, generally speaking, not finite in the spatial variables x and y; $g(z)$ is the generalized function of the order N with support in the point $z = H$

$$g(z) = \sum_{k=0}^{N} c_k \delta^{(k)}(z - H).$$

Here, c_k are certain constants, δ_k is the k-derivative of the Dirac function.

Consider the inverse problem: to determine $f(x, y, t)$ for known parameters α, β, N, H, and c_k by the trace of solution $u(x, y, z, t) \in S'(R^4)$ of (1.61) in the plane $z = 0$

$$u(x, y, z, t)\Big|_{z=0} = u_0(x, y, t). \tag{5.2}$$

The function $f(x, y, t)g(z)$ may be interpreted as the wave radiation source; the function $u_0(x, y, t)$, as the field generated by this source; the constants α, β characterize the field damping.

The basic result. Before we formulate the theorem, we must show that for a fixed $f(x, y, t) \in S'(R^3)$ the solution $u(x, y, z, t)$ to (1.61) from $S'(R^4)$ exists, is unique, and, besides, for functions $g(z)$ of a certain form, the procedure of taking the field trace is correct. Thus, the following lemma holds.

Lemma 1.5. Let $f(x, y, t) \in S'(R^3)$ and $g(z) = \sum\limits_{k=0}^{N} c_k \delta^{(k)}(z - H)$, where $N = 0$ for $H = 0$. Then the solution to (1.61) exists in $S'(R^4)$, is unique, and belongs to $C^\infty(R^1/\{H\}) \bigcap C(R^1)$ by the variable z.

Proof. Use the properties of the Fourier transform in $S'(R^4)$. Suppose the following relations holds:

$$\hat{\varphi}(\lambda, \mu, \xi, \omega) = \int\limits_{R^4} \varphi(x, y, z, t) e^{-i(\lambda x + \mu y + \xi z, -\omega t)} dx dy dz dt,$$

where $\varphi \in S(R^4)$.

Apply it to a homogeneous equation of the form (1.61)

$$\hat{u}(\lambda, \mu, \xi, \omega)\Big[-\lambda^2 - \mu^2 - \xi^2 + \frac{\omega^2}{c} + i\beta\omega - \alpha^2\Big] = 0. \qquad (5.3)$$

As the polynomial in spatial-time frequences in (1.63) has no real roots, then by the Hörmander-Loyasevich theorem (Vladimirov, 1976) the solution to (1.61) in $S'(R^4)$ exists for each right side from $S'(R^4)$. As the Fourier transformation in this space is univalent, it is unique. To find it, apply to (1.61) the Fourier transformation by x, y and t

$$\frac{\partial^2 u'}{\partial z^2} - su' = -f'(\lambda, \mu, \omega) \sum_{k=0}^{N} c_k \delta^{(k)}(z - H). \qquad (5.4)$$

Here $u' = u'(z, \lambda, \mu, \omega)$; $s = \lambda^2 + \mu^2 - \omega^2/c^2 - i\beta\omega + \alpha^2$; $\hat{f}(\lambda, \mu, \omega)$ is the Fourier image of $f(x, y, t)$.

It is easy to show that the function

$$u'(z, \lambda, \mu, \omega) = \frac{\hat{f}(\lambda, \mu, \omega)}{2} \sum_{k=0}^{N} c_k \frac{\partial^k}{\partial z^k} \left(\frac{e^{-|z-H|\sqrt{s}}}{\sqrt{s}} \right). \qquad (5.5)$$

satisfy this differential equation.

As the unique solution to (1.61) belongs to $S'(R^4)$, then its Fourier transform by x, y and t also belongs to $S'(R^4)$ (Vladimirov, 1976). But this holds only in the case when $\mathrm{Re}(\sqrt{s}) > 0$ (when $\mathrm{Re}(\sqrt{s}) < 0$, the function $u'(z, \lambda, \mu, \omega)$ exponentially grows in infinity, and therefore, does not belong to $S'(R^4)$). The sign of $\mathrm{Im}(\sqrt{s})$ in this case is determined from the relation

$$\mathrm{Re}(\sqrt{s}) \cdot \mathrm{Im}(\sqrt{s}) = -\beta\omega.$$

The obtained Fourier transform, as it is easy to see, belongs to $C^\infty(R^1/\{H\})$ by z. Thus, for $H \neq 0$, the function $u'(z, \lambda, \mu, \omega)$ trace in the point $z = 0$ is determined correctly

$$u'(0, \lambda, \mu, \omega) = \hat{u}_0(\lambda, \mu, \omega) \in S'(R^3).$$

If $H = 0$, such operation may be done only for $g(z)$ being the zero order function $c_0\delta(z)$, since in the general case, in the point $z = 0$, the first derivative in (1.65) is undetermined and the others are singular. Thus, for the chosen function $g(z)$, we have $\hat{u}_0(\lambda,\mu,\omega) \in S'(R^3)$ and $f(x,y,t) \in S'(R^3)$. We shall prove the following theorem using this lemma.

Theorem 1.7. Let $g(z) = \sum\limits_{k=0}^{N} c_k\delta^{(k)}(z-H)$ for $H \neq 0$ and $g(z) = c_0\delta(z)$ for $H = 0$. Then for $f(x,y,t) \in S'(R^3)$ the inversion formula

$$f(x,y,t) = \int\limits_{R^3} \hat{u}_0(\lambda,\mu,\omega)R(\lambda,\mu,\omega;H)e^{i(\lambda x+\mu y-\omega t)}d\lambda d\mu d\omega \qquad (5.6)$$

holds. Here $\hat{u}_0(\lambda,\mu,\omega)$ is the Fourier image of $u_0(x,y,t)$;

$$R(\lambda,\mu,\omega;H) = 2e^{|H|\sqrt{s}}\left(\sum_{k=0}^{N} s^{\frac{k-1}{2}}\left(-\mathrm{sign}(H)\right)^k\right)^{-1}, \qquad (5.7)$$

$$s = \lambda^2 + \mu^2 - \frac{\omega^2}{c^2} - i\beta\omega + \alpha^2; \quad \mathrm{Re}(\sqrt{s}) > 0.$$

Proof. The existence and uniqueness of the solution to such a problem follows from univalent representation of the field spectrum in the plane $z = 0$ *via* the source $f(x,y,t)$ spectrum which, on the basis of (1.65), has the form

$$u_0(\lambda,\mu,\omega) = \frac{1}{2}\hat{f}(\lambda,\mu,\omega)e^{-|H|\sqrt{s}}\sum_{k=0}^{N} c_k s^{\frac{k-1}{2}}\left(-\mathrm{sign}(H)\right)^k. \qquad (5.8)$$

Substituting (1.68) by $R(\lambda,\mu,\omega;H)$ determined in (1.67) and doing the inverse Fourier transformation, we obtain the required relation (1.66).

Remark 1. Note that the obtained solution continuously depends on $u_0(x,y,t)$. Really, let $u_0^n(x,y,t) \to 0$ as $n \to \infty$ in the sense of $S'(R^3)$ convergence. As the Fourier transformation is continuous in $S'(R)$, then $u_0^n(\lambda,\mu,t) \to 0$ as $n \to \infty$. According as (1.68), $\hat{f}^n(\lambda,\mu,\omega) \to 0$ as $n \to \infty$. Hence $f^n(x,y,t) \to 0$ as $n \to \infty$.

Remark 2. Since the function $1/R(\lambda,\mu,\omega;H)$ belongs to the multiplicator set in $S(R^3)$, then the analogous statement holds when $f(x,y,t) \in S(R^3)$. In this case, $u_0(x,y,t)$ will belong to $S(R^3)$ and the invertion formula (1.66) holds.

Remark 3. Solving the two-dimensional problem (1.61)-(1.62) leads to formulas analogous to (1.66)-(1.68). The only difference is the absence in the formulas of the variables y and μ.

Asymptotic representation of solution. Suppose that in the two-dimensional case, the field trace in the line $z=0$ is as follows

$$u_0(x,t) = \varphi(x - vt), \tag{5.9}$$

where $v < c$ and $\varphi(\tau)$ is a real even function. Then

$$f(x,t) = \psi(x - vt), \tag{5.10}$$

where $\psi(\tau)$ is also a real even function. As $\alpha \to 0$, $\beta \to 0$, $g(z) = \delta'(z - H)$, $H \neq 0$ the representation

$$\psi(\tau) = 2\sum_{k=0}^{\infty}(-1)^k \frac{H^{2k}(1-\gamma^2)^k}{(2k)!}\frac{\partial^{2k}\varphi(\tau)}{\partial\tau^{2k}}. \tag{5.11}$$

holds. Here $\gamma = v/c$ and $\tau = x - vt$.

Really, the field spectrum in the line $z=0$ in this case is determined by $\hat{u}_0(\lambda,\omega) = \hat{v}p(\lambda)\delta(\omega - \lambda v)$. Then, according as (1.66),

$$f(x,t) = \int_{R^1} \hat{\varphi}(\lambda)R(\lambda,\lambda v;H)e^{i\lambda(x-vt)}d\lambda, \tag{5.12}$$

and, thus, $f(x,t) = \psi(x - vt)$. This means that the source generated field (1.69) moves with the speed v along the axis x. As $\varphi(\tau)$ is real and even, its spectrum will be same

$$\varphi(\tau) = 2\int_0^\infty \hat{\varphi}(\lambda)\cos\lambda\tau d\tau. \tag{5.13}$$

Taking into account (1.72), we obtain the equality as $\alpha \to 0, \beta \to 0$

$$\begin{aligned}
\psi(\tau) & \int_{-\infty}^{+\infty} \hat{\varphi}(\lambda) R(\lambda, \lambda v; H) e^{i\lambda\tau} d\lambda \\
&= \int_{-\infty}^{+\infty} \hat{\varphi}(\lambda) R(\lambda, \lambda v; H) e^{|\lambda||H|\sqrt{1-\gamma^2}} e^{i\lambda\tau} d\lambda \\
&= \int_0^\infty \hat{\varphi}(\lambda) e^{|H|\lambda\sqrt{1-\gamma^2}} \cos\lambda\tau d\tau.
\end{aligned} \tag{5.14}$$

Taking into account that for $\lambda \gg 1$, we have the asymptotic representation

$$e^{|H|\lambda\sqrt{1-\gamma^2}} \approx 2\mathrm{ch}\left(H\lambda\sqrt{1-\gamma^2}\right) = 2\sum_{k=0}^{\infty} \frac{\lambda^{2k} H^{2k}}{(2k)!}(1-\gamma^2)^k,$$

and comparing (1.73) with (1.74), we obtain the source function $\psi(\tau)$ as the result of applying a certain differential operator of infinite order (formula (1.71)) to the initial information $\varphi(\tau)$.

Applications. Below, we shall give the explicit formulas of the trace of (1.61) solution for some sources of the type "moving oscillating point". Certain peculiarities of behavior of the solutions obtained are investigated. The calculations are done both for two-dimensional and three-dimensional cases. Further, for definiteness, assume that $\beta > 0$.

In the line $z = 0$, to the wave field of the type

$$u_0(x,t) = \frac{1}{\sqrt{1-\gamma^2}} \exp\left(\frac{-i\omega_0}{\sqrt{1-\gamma^2}}\left(t - \frac{v}{c^2}p\right)\right)$$

$$\times \exp\left(-\frac{\beta v}{2}\frac{p}{\sqrt{1-\gamma^2}}\right) K_0(kr) \qquad (5.15)$$

there corresponds a source of the type

$$f(x,t)g(z) = e^{-i\omega_0 t}\delta(x - vt)\delta(z - H).$$

Here $\gamma = v/c, \gamma < 1, K_0$ is the Mcdonal'd function of zero order;

$$\omega_0 > 0; \quad p = (x - vt)/\sqrt{1-\gamma^2}; \quad r = (p^2 + H^2)^{1/2},$$

$$k^2 = \frac{1}{1-\gamma^2}\left(\frac{\beta^2 v^2}{4} + \alpha^2(1-\gamma^2) - \frac{\omega_0^2}{c^2} - i\beta\omega_0\right), \quad \mathrm{Re}(k) > 0. \quad (5.16)$$

Expression (1.75) is obtained by means of (1.68), where the spectrum of the source function is substituted (in our case $\hat{f}(\lambda,\omega) = \hat{\delta}(\omega - \lambda v - \omega_0)$). Calculating the inverse Fourier transformation, after the substitution

$$\lambda' = \lambda\sqrt{1-\gamma^2} - \frac{1}{2\sqrt{1-\gamma^2}}\left(\frac{2v\omega_0}{c^2} + 2\beta v\right)$$

we may show that the integration line may be transposed to the real axis. Then, on the basis of (7.59) from Ditkin and Prudnikov (1974), we obtain (1.75).

Note, that if $v \to 0, \alpha \to 0, \beta \to 0$, expression (1.75) is transformed to the known formula for a harmonic source

$$u_0(x,t) = e^{i\omega_0 t} H_0^{(1)}\left(\frac{\omega_0}{c}\sqrt{x^2 + H^2}\right).$$

Consider field (1.75) in the point $x = 0$. Taking account of the asymptotic representation of the Mcdonal'd function

$$K_0(z) \sim e^{-z}\sqrt{\frac{2z}{\pi}}, \qquad z \gg 1,$$

we observe the Doppler effect of changing the instantaneous frequency of the carrying signal

$$\omega_{\text{dop}}(t) = \frac{\omega_0}{1-\gamma^2}\left(1 - \frac{v\gamma t}{\sqrt{v^2t^2 + H^2(1-\gamma^2)}}\right)$$

The signal envelope, which is obtained from (1.75) for $\omega_0 = 0$, is as follows

$$u_0(0,t) = \frac{\exp\left(\frac{\beta v^2 t}{2(1-\gamma^2)}\right)}{\sqrt{1-\gamma^2}} K_0\left(\sqrt{\left(\frac{\beta^2 v^2}{4(1-\gamma^2)} + \alpha^2\right)\left(\frac{v^2t^2}{1-\gamma^2} + H^2\right)}\right), \tag{5.17}$$

whence for $-t \gg 1$ and $\alpha^2 \ll 1$, we obtain

$$u_0(0,t) = \frac{\exp\left(\frac{\beta v^2 t}{1-\gamma^2}\right)}{\sqrt{v|t|}}. \tag{5.18}$$

Thus, in dissipative media, the field of a moving source has, apart from ordinary weakening $1/\sqrt{v|t|}$, additional exponential damping that depends on the speed with which the source is moving.

In the plane $z = 0$, to the oscillating field

$$u_0(x,y,t) = \frac{1}{\sqrt{1-\gamma^2}} \exp\left(\frac{-i\omega_0}{\sqrt{1-\gamma^2}}\left(t - \frac{v}{c^2}p\right)\right) \times \exp\left(-\frac{\beta v}{2}\frac{p}{\sqrt{1-\gamma^2}}\right)\frac{\exp(kR)}{R} \tag{5.19}$$

there corresponds the source

$$f(x,y,t)g(z) = e^{-i\omega_0 t}\delta(x - vt)\delta(y)\delta(z - H).$$

Here, as before, $\omega_0 > 0$, k is determined by (1.76), and $R = (p^2 + y^2 + H^2)^{1/2}$.

The field (1.79) trace may be calculated according as (6.646) from Gradshteyn and Ryzhik (1971) after passing for inverse Fourier transformation into the complex plane and introducting the polar coordinate system.

As in the previous case, here we have the Doppler effect, and as $v \to 0, \alpha \to 0, \beta \to 0$, formula (1.79) gives the known solution for a harmonic source.

$$u_0(x,y,t) = \frac{\exp\left(-i\omega_0\left(t - \frac{\sqrt{x^2+y^2+H^2}}{c}\right)\right)}{\sqrt{x^2+y^2+H^2}}.$$

Similar to (1.77), the signal envelope in the line $x = 0$ has the form

$$u_0(0,y,t) = \frac{\exp\left(\frac{\beta v^2 t}{2(1-\gamma^2)}\right)}{\sqrt{1-\gamma^2}} \frac{\exp\left(\frac{\beta v}{2\sqrt{1-\gamma^2}}\sqrt{\frac{v^2t^2}{1-\gamma^2} + y^2 + H^2}\right)}{\sqrt{\frac{v^2t^2}{1-\gamma^2} + y^2 + H^2}}. \tag{5.20}$$

Hence, the side damping of the signal in the plane $z = 0$ is as follows

$$u_0(0,y,t) = \frac{1}{\sqrt{1-\gamma^2}} \frac{\exp\left(\frac{\beta v}{2\sqrt{1-\gamma^2}}\sqrt{y^2+H^2}\right)}{\sqrt{y^2+H^2}}.$$

For $y = 0, -vt \gg H$, and taking into account (1.80), we obtain an expression analogous to (1.78)

$$u_0(0,0,t) = \frac{\exp\left(\frac{\beta v^2}{1-\gamma^2}t\right)}{v|t|}.$$

In the line $z = 0$ to the wave field

$$u_0(x,t) = \frac{\exp\left(-\frac{i\omega_0}{\sqrt{1-\gamma^2}}\left(t - \frac{v}{c^2}p\right) - \frac{\beta\nu p}{2\sqrt{1-\gamma^2}}\right)}{\sqrt{1-\gamma^2}} \sum_{n=0}^{N} c_n \frac{\partial^n}{\partial H^n} K_0(kr) \tag{5.21}$$

there corresponds the source

$$f(x,t)g(z) = e^{-i\omega_0 t}\delta(x - vt)\sum_{n=0}^{N} c_n \delta^{(n)}(z - H).$$

Here $N = 0$ for $H = 0$ and the derivatives of the Mcdonal'd function of zero order are calculated according as (8.486) from Gradshteyn and Ryzhik (1971) by the recurrence formula

$$\frac{d}{dz}K_\nu(z) = -\frac{\nu}{z}K_\nu(z) - K_{\nu+1}(z), \quad \frac{d}{dz}K_0(z) = -K_1(z), \quad \nu = 1, 2, \ldots$$

Really, from the initial formulas (1.65), (1.68), we may see that the field trace from the source $f(x,t)\delta^{(n)}(z-H)$ is calculated by formal differentiating n times by H of the source field $f(x,t)\delta(z-H)$. Then, on the basis of (1.75), we obtain (1.81).

Taking into account the obtained formula we may write out the solution for dipole

$$f(x,t)g(z) = e^{-i\omega_0 t}\delta(x - vt)\delta'(z - H).$$

In this case

$$u_0(x,t) = \frac{kH\exp\left(\frac{-i\omega_0}{\sqrt{1-\gamma^2}}\left(t - \frac{v}{c^2}p\right)\right)}{4r\sqrt{1-\gamma^2}} \cdot \exp\left(\frac{\beta vp}{2\sqrt{1-\gamma^2}}\right) K_1(kr).$$

Unlike the field for monopole type source (formula (1.75)), there arises here a factor of the form

$$\frac{H}{\sqrt{(x-vt)^2/(1-\gamma^2) + H^2}} = \cos\left(\operatorname{arctg}\frac{H\sqrt{1-\gamma^2}}{x - vt}\right)$$

which characterizes the nonisotropy inherent in a dipole source.

Remark. In the three-dimensional case, differentiating by H the right side of (1.79), we may obtain an analogous formula for the sources

$$f(x,y,t)g(z) = e^{i\omega_0 t}\delta(x - vt)\delta(y)\sum_{n=0}^{N} c_n\delta^{(n)}(z - H).$$

In the line $z = 0$, to the field

$$u_0(x,t) = \frac{i \exp\left(\frac{-i\omega_0}{1-\gamma^2}\left(t - \frac{v}{c^2}p\right)\right)}{\sqrt{1-\gamma^2}} \exp\left(\frac{\beta v p}{2\sqrt{1-\gamma^2}}\right)$$

$$\times \left[\frac{1}{2\sqrt{1-\gamma^2}}\left(\frac{2\omega_0}{c^2} + i\beta v\right) K_0(kr) + 2 \sin\left(\operatorname{arctg}\frac{p}{H}\right) K_1(kr)\right] \quad (5.22)$$

there corresponds a source of the type

$$f(x,t)g(z) = e^{-i\omega_0 t}\delta'(x - vt)\delta(z - H).$$

Here, as earlier, $p = (x - vt)/\sqrt{1-\gamma^2}$, $r = (p^2 + H^2)^{1/2}$.

Field representation (1.82) is obtained by the formula (8.67) from Ditkin and Prudnikov (1974). In (1.82) the factor $\sin(\operatorname{arctg} p/H)$ characterizes direction of the source. Unlike (1.81), the intensity maximum of radiation here coincides with the direction of motion.

In the three-dimensional case, to the field

$$u_0(x,y,t) = \frac{\exp\left(\frac{-i\omega_0}{\sqrt{1-\gamma^2}}\left(t - \frac{v}{c^2}p\right)\right)}{\sqrt{1-\gamma^2}} \exp\left(\frac{-\beta v p}{2\sqrt{1-\gamma^2}}\right)$$

$$\times \left[\frac{1}{2\sqrt{1-\gamma^2}}\left(\frac{2i\omega_0}{c^2}v\right)\frac{e^{kR}}{R} + \frac{4}{k}\sqrt{2\pi}\cos\left(\operatorname{arctg}\frac{y}{p}\right)\frac{\sqrt{y^2+p^2}}{R^{3/2}} K_{3/2}(kR)\right] \quad (5.23)$$

there corresponds the source

$$f(x,y,t)g(z) = e^{-i\omega_0 t}\delta'(x - vt)\delta(y)\delta(z - H).$$

Here, also, $p = (x - vt)/\sqrt{1-\gamma^2}$, $R = (p^2 + y^2 + H^2)^{1/2}$, $\gamma = v/c$.

Expression (1.83) is obtained by formula (6.645) from Gradshteyn and Ryzhik (1971). As in the two-dimensional case, the direction of the source is determined by the factor

$$\cos\left(\operatorname{arctg}\frac{y}{p}\right)\frac{\sqrt{y^2+p^2}}{R} = \cos\left(\operatorname{arctg}\frac{y\sqrt{1-\gamma^2}}{x - vt}\right)\cos\left(\operatorname{arctg}\frac{H}{\sqrt{\frac{(x-vt)^2}{1-\gamma^2} + y^2}}\right).$$

In the plane $z = 0$ to the field

$$u_0(x, y, t) = \frac{4\sqrt{2\pi}}{\sqrt{k(1-\gamma^2)}} \sin\left(\operatorname{arctg}\frac{y}{p}\right) \frac{\sqrt{y^2+p^2}}{R^{3/2}}$$

$$\times \exp\left(\frac{-i\omega_0}{\sqrt{1-\gamma^2}}\left(t - \frac{v}{c^2}p\right)\right) \exp\left(-\frac{\beta v}{2}\frac{p}{\sqrt{1-\gamma^2}}\right) K_{3/2}(kR) \quad (5.24)$$

there corresponds the anisotropic source

$$f(x, y, t)g(z) = e^{-i\omega_0 t}\delta(x - vt)\delta'(y)\delta(z - H).$$

Here, also, $p = (x - vt)/\sqrt{1-\gamma^2}$, $R = (p^2 + y^2 + H^2)^{1/2}$.

Formula (1.84) is obtained by (6.645) from Gradshteyn and Ryzhik (1971). The source direction is characterized by the factor

$$\sin\left(\frac{\operatorname{arctg} y\sqrt{1-\gamma^2}}{x - vt}\right) \cos\left(\frac{\operatorname{arctg} H\sqrt{1-\gamma^2}}{\sqrt{(x-vt)^2 + y^2(1-\gamma^2)}}\right).$$

This formula shows that the direction of maximal intensity of source radiation is perpendicular to the trajectory of motion and lies in the plane $z = H$.

The problem for the dispersion equation. Consider the equation

$$\Delta u - \frac{1}{c^2}\frac{\partial^2 u}{\partial t^2} + q^2\frac{\partial}{\partial t}\Delta u - \beta\frac{\partial u}{\partial t} - \alpha^2 u = -f(x, y, t)g(z), \quad (5.25)$$

where $x, y, z, t \in R^1$; $q \neq 0$, $\beta \neq 0$, $\alpha \neq 0$; $f(x, y, t) \in S'(R^3)$, $g(z) = \sum_{k=0}^{N} c_k\delta^{(k)}(z - H)$. Here q^2 is the dissipative coefficient, which characterizes the viscosity and heat conductivity of the medium (Vinogradov *et al*,1979).

As before, we need to determine $f(x, y, t)$, knowing the coefficients of (1.85), the function $g(z)$ and the solution $u(x, y, z, t) \in S'(R^4)$ trace in the plane $z = 0$

$$u(x,y,z,t)\Big|_{z=0} = u_0(x,y,t). \tag{5.26}$$

A theorem analogous to theorem 1.7 holds.

Theorem 1.8. Let $\beta > 0$ and $N = 0$ for $H = 0$. Then a solution to inverse problem (1.85)-(1.86) exists, is unique and

$$f(x,y,t) = \int\limits_{R^3} \hat{u}_0(\lambda,\mu,\omega) R(\lambda,\mu,\omega;H) e^{i(\lambda x+\mu y-\omega t)} d\lambda d\mu d\omega,$$

where $\hat{u}_0(\lambda,\mu,\omega)$ is the Fourier image of $u_0(x,y,t) \in S'(R^3)$;

$$R(\lambda,\mu,\omega;H) = 2\left(1 - iq^2\omega\right) e^{|H|\sqrt{s_1}} \left(\sum_{k=0}^{N} c_k s_1^{\frac{k-1}{2}} \left(-\mathrm{sign}(H)\right)^k\right)^{-1};$$

$$s_1 = \lambda^2 + \mu^2 + \frac{\alpha^2 - \omega^2/c^2 - i\beta\omega}{1 - iq^2\omega}; \quad \mathrm{Re}(\sqrt{s_1}) > 0.$$

The proof of this theorem is analogous to those for problem (1.61)-(1.62). We shall show only the necessity of a restriction on the coefficient for uniqueness of the solution to equation (1.85) in $S'(R^4)$. For this purpose, apply to homogeneous equation of the form (1.85) the Fourier transformation by variables x, y, z, t

$$\hat{u}(\lambda,\mu,\xi,\omega)\left[(\lambda^2+\mu^2+\xi^2)(1-iq^2\omega) - \frac{\omega^2}{c^2} - i\beta\omega + \alpha^2\right] = 0. \tag{5.27}$$

We see that the real part of the expression in square brackets for $\omega = 0$ is always not zero for each real λ, μ, and ξ. For $\omega \neq 0$, real and imaginary parts may be equal to zero simultaneously only if $\beta < 0$. Thus, for $\beta > 0$, from (1.87) it follows that $\hat{u}(\lambda,\mu,\xi,\omega) \equiv 0$. The property of Fourier transformation in $S'(R^4)$ yields that $u(x,y,z,t) = 0$. Thus, the solution to (1.87) in the space $S'(R^4)$ is unique for each right side from this space.

The problems for parabolic and elliptic equations. The following method allows us to consider the problems (1.61)-(1.62) for other types of differential equations, obtained on the basis of equation (1.61). Theorem 1.7 and the inversion formula analogous to (1.66) also hold true for them. We shall give three such settings. As earlier,

$$g(z) = \sum_{k=0}^{N} c_k \delta^{(k)}(z - H); \quad N = 0 \quad \text{for} \quad H = 0.$$

Denote $a = 1/c^2$. Then, formally setting $a = 0, \beta > 0$ in (1.61), we obtain the parabolic equation

$$\Delta u - \beta \frac{\partial u}{\partial t} - \alpha^2 u = -f(x, y, t) g(z),$$

where $x, y, z \in R^1, \alpha \neq 0, \beta > 0, f(x, y, t) \in S'(R^3)$.

The inverse problem is to determine $f(x, y, t)$ by the trace of solution $u(x, y, z, t) \in S'(R^4)$ in the plane $z = 0$

$$u(x, y, z, t)\Big|_{z=0} = u_0(x, y, t).$$

It may be shown that in the inversion formula (1.66) in this case $s = \lambda^2 + \mu^2 - i\beta\omega + \alpha^2$ and the above formulas hold.

Further, let $a = 0, \beta = 0$ and suppose that the solution to (1.61) does not depend on time. Then, this equation becomes the elliptic type equation

$$\Delta u - \alpha^2 u = f(x, y) g(z),$$

where $x, y, z \in R^1, \alpha \neq 0, f(x, y) \in S'(R^2); u(x, y, z) \in S'(R^3)$. The inverse problem is to determine $f(x, y)$ by the solution trace in the plane $z = 0$

$$u(x, y, z)\Big|_{z=0} = u_0(x, y).$$

In this case, in the inversion formula (1.66), we must set $s = \lambda^2 + \mu^2 + \alpha^2$.

Apply to (1.61)-(1.62) the Fourier transformation by time and set $\beta = 2\alpha/c$. Then (1.61) becomes the Helmholtz equation with a complex wave number and (1.61)-(1.62) takes the form

$$\Delta u + k^2 u = -f(x, y, \omega) g(z),$$

$$u(x, y, z, \omega)\Big|_{z=0} = u_0(x, y, \omega).$$

Here $x, y, z, \omega \in R^1$, $k = \frac{\omega}{c} + i\alpha$; $f(x, y, \omega) \in S'(R^3)$.

The inversion formula then becomes as follows

$$f(x, y, \omega) = \int_{R^2} \hat{u}_0(\lambda, \mu, \omega) R(\lambda, \mu, \omega, H) e^{i(\lambda x + \mu y)} dx dy,$$

where $R(\lambda, \mu, \omega, H)$ is determined by (1.67).

§1.6 Conjugated equations in control problems of nonstationary geophysical fields.

Conjugate equations and the perturbation algorithm method are widely used to consruct problem-solving algorithms (Marchuk *et al.*,1986). We shall consider the conjugate equations method for problems of geophysical fields control on the basis of variation of radiation source parameters. For this purpose, first, consider the following simplest setting of the problem.

Let us have N seismic or acoustic sources, the radiation parameters of which may be changed. It is required to pose their radiation such that, in a certain small neighborhood of $x_0 \in R^3$, the pressure or the shift attains its maximal modulus value in a fixed moment $t_0 \in R^1$.

Mathematically, such setting is as follows. Consider the equation

$$Lu = F(x, t), \tag{6.1}$$

where L is a certain differential operator, for example, the wave operator

$$L = \frac{1}{c}\frac{\partial^2}{\partial t^2} - \Delta; \tag{6.2}$$

the function $F(x,t)$, $x \in R^3$, $t \in R^1$ describes the sources

$$F(x,t) = \sum_{k=1}^{N} f_k(t)\delta(x - x^k), \tag{6.3}$$

and the function $u(x,t)$ characterizes the oscillation field in the point $x \in R^3$. Assume that the source's cooordinates are fixed and known. Then the problem may be formulated as the problem of maximizing the functional

$$J_g(u) = (u,g) = \int_{R^4} u(x,t)g(x,t)dxdt, \tag{6.4}$$

where $g(x,t) = \delta(t-t_0)\delta(x-x_0)$ for certain restrictions on the function $f_k(t)$.

Before we solve the set problem we must consider some auxiliary definitions and statements.

Let

$$P(y, D_y) = \sum_{|\alpha|\le k} c_\alpha(y) D_y^\alpha$$

be a differential operator in the space $D(R^{n+1})$.

The operator

$$P^*(y, D_y) = \sum_{|\alpha|\le k} (-1)^{|\alpha|} D_y^\alpha c_\alpha(y)$$

is formally a Lagrange conjugate to the operator $P(y, D_y)$. In this case, we have (Tikhonov *et al.*,1983)

$$(Pu, v) = (u, P^*v) \quad \forall u, v \in D(R^{n+1}), \tag{6.5}$$

where (,) is the scalar Hilbert product in R^{n+1}.

Consider the hyperbolic equation

$$Pu = u_{tt} - Au = f(x,t), \tag{6.6}$$

where A is an elliptic self-conjugate second order operator, $x \in D \subset R^n$, $t \in [0,T] \subset R^1$.

For equation (1.93), the trivial boundary conditions

$$u\Big|_{\partial D} = 0 \tag{6.7}$$

and nontrivial initial data

$$u\Big|_{t=0} = \varphi_0(x), \tag{6.8}$$

$$u_t\Big|_{t=0} = \varphi_1(x) \tag{6.9}$$

are given. For the conjugate equation, the analogous setting will be as follows

$$P^*v = v_{tt} - A^*v = g(x,t), \tag{6.10}$$

$$v\Big|_{\partial D} = 0, \tag{6.11}$$

$$v\Big|_{t=T} = \psi_0(x), \tag{6.12}$$

$$v_t\Big|_{t=T} = \psi_1(x). \tag{6.13}$$

Lemma 1.6. Let problems (1.93)-(1.96) and (1.97)-(1.100) have a unique solution in the class $D(R^{n+1})$ and $v, g \in D(R^{n+1})$. Then, in order that (1.92) holds, it is sufficient that

$$(f, v) = (g, u), \tag{6.14}$$

$$\varphi_0(x) = \varphi_1(x) = \psi_0(x) = \psi_1(x) = 0. \tag{6.15}$$

Proof. Multiply equation (1.93) by $v(x,t)$ and equation (1.97) by $u(x,t)$ in a scalar way and subtract them from each other

$$Q = (Pu, v) - (u, P^*v).$$

Integrating term-by-term the first scalar product, we obtain

$$Q = (f, v) - (g, u) + u_t(T, x)\psi_0(x) - \varphi_1(x)v(0, x)$$

$$-u(T, x)\psi_0(x) + \varphi_0(x)v(0, x).$$

In order that $Q = 0$, it is sufficient that $(f, v) = (g, u)$ for $v, g \in D(R^{n+1})$ and $\varphi_0(x) = \varphi_1(x), \psi_0(x) = 0$.

Analogously, using the second scalar product, we obtain $(f, v) = (g, u)$ and $\psi_0(x) = \psi_1(x), \varphi_0(x) = 0$, which was to be proved.

Remark. We now investigate the structure of fundamental solutions of direct (1.93)-(1.96) and conjugate (1.97)-(1.100) problems for the wave equation in the three-dimensional case under conditions of lemma 1.6. The speed in the medium is assumed to be constant and equal to c.

The fundamental solution will be as follows

$$\mathcal{E}_-(x,t) = \frac{\Theta_-(t)\delta(|x| - ct)}{4\pi a|x|}, \quad \Theta_-(t) = \begin{cases} 1, & t \geq 0, \\ 0, & t < 0, \end{cases} \tag{6.16}$$

where $|x| = (x_1^2 + x_2^2 + x_3^2)^{1/2}$. The right side is represented by way of asymmetric generalized functions

$$f(x,t) = \delta_-(t)\delta(x).$$

Here $(\delta_-(t), \varphi(t)) = \varphi(0+) \ \ \forall \varphi \in D(R^1)$. The time in (1.93)-(1.96) is changed from 0 to T.

For the conjugate problem, the fundamental solution is such

$$\mathcal{E}_+(x,t) = \frac{\Theta_+(t-T)\delta(|x| + c(t-T))}{4\pi a|x|}, \quad \Theta_+(t) = \begin{cases} 1, & t \leq 0, \\ 0, & t > 0. \end{cases} \tag{6.17}$$

The corresponding right side of (1.97) is

$$g(x,t) = \delta_+(t-T)\delta(x),$$

where $(\delta_+(t), \varphi(t)) = \varphi(0-)\ \forall\varphi \in D(R^1)$. The time for the conjugate problem is changed from T to 0.

Now we can return to the initial control problem (1.88)-(1.91). Let conditions of lemma 1.6 hold. Then, according to (1.101), the problem of maximizing functional (1.91) is equivalent to the analogous problem for the functional (v, F)

$$J_g(u) = (u, g) = (v, F).$$

Here $F(x,t)$ is determined by (1.90); $v(x,t)$ solves conjugate problem (1.97)-(1.100) with trivial initial data for the right side $g(x,t) = \delta(x - x_0)\delta(t - t_0)$. Taking account of the remark to lemma 1.6 (formula (1.104)), the solution $v(x,t)$ will be the following

$$v(x,t) = \frac{\Theta_+(t-t_0)\delta\big(|x-x_0| + c(t-t_0)\big)}{4\pi a|x-x_0|}.$$

Then

$$J_g(u) = \sum_{k=1}^{N} \int_0^{t_0} \Theta_+(t-t_0)\frac{\delta\big(t - t_0 + |x-x_0|/c\big)}{4\pi c^2|x_k - x_0|} f_k(t)dt$$

$$= \sum_{k=1}^{N} \frac{f_k\big(t_0 - |x_k - x_0|/c\big)}{4\pi c^2 |x_k - x_0|}. \tag{6.18}$$

Assume, for example, that $f_k(t) > 0$, $f_k(t) = \varphi(t - \tau_k)$ and $\varepsilon = \max_{t\in R^1} |\varphi(t)|$. Then the maximum of (1.91) is attained when

$$t_0 - \frac{|x_k - x_0|}{c} - \tau_k = \varepsilon.$$

Hence, the delay times τ_k for the sources are

$$\tau_k = t_0 - \varepsilon - \frac{|x_k - x_0|}{c}, \qquad k = 1, 2, \ldots, N. \tag{6.19}$$

Remark. Analogous methods may be applied for different problem settings for source control; for example, for problems where we must control source parameters in order to ensure that the field (or its functionals) would not exceed given values c_i in certain domains $D_i \subset R^3$, $i = 1, \dots, M$. Such settings are widely used, for example, in meteorology in problems of requiring preservation of the surrounding medium (Marchuk, 1976; Marchuk *et al.*,1986). For seismology, such problems may be considered as those modelling variable seismic processes in the Earth's interior.

§1.7 Inverse problem of source determination as the problem of statistical evaluation of parameters of the mathematical model.

We shall consider the general approach to passive location as solving the inverse problem by the methods of determining statistical estimates of the parameters of the desired sources and, possibly, those of the medium as well.

Setting of the inverse problem. Consider the system of equations

$$\mathcal{L}\bar{u} = \bar{f}(x, t), \tag{7.1}$$

where $x \in D \subseteq R^n, t \in [0, T], \mathcal{L} = \mathcal{L}(x, D_x, \partial/\partial t)$ is a matrix $N \times N$ integro-differential (or pseudodifferential) operator of the order $m \leq 2$ describing the physical field propagation $u(x, t) = (u_1, \dots, u_n)$ in a medium. Let $Q = D \times [0, T]$, $q \in Q$, $q = (x, t)$.

Assume that, for equation (1.107), the corresponding initial and boundary conditions are given. For a hyperbolic equation, let them be as follows

$$Bu\Big|_{\partial D} = \varphi, \tag{7.2}$$

$$u\Big|_{t=0} = 0, \tag{7.3}$$

$$u_t\Big|_{t=0} = 0. \tag{7.4}$$

Here B is an operator acting on ∂D of the order $m_B < 2$. Let

$$\mathcal{L}(x, D_x, \partial/\partial t) = \mathcal{L}(c(x), D_x, \partial/\partial t),$$

where $c(x) = \big(c_1(x), \ldots, c_{L(x)}\big)$ is the coefficient vector of $\mathcal{L}$. Denote by $d(q)$ the vector function consisting of the right side coefficients of (1.107)

$$d(q) = \big(c(x), f(x,t)\big).$$

The inverse problem is to find the vector function $d(q)$ by means of the information

$$v(r,t) = u(r,t;d) + \varepsilon(r,t). \tag{7.5}$$

Here $r \in R \subset D$, R being a smooth manifold in D where the information is given; $\varepsilon(r,t)$ is noise of normal distribution with zero average and known covariance matrix $\Gamma_\varepsilon(q,q')$; $q, q' \in R$.

Denote by H_B the Hilbert weight space of vector functions $P(q) = \big(p_1(q), \ldots, p_l(q)\big)$ with the scalar product

$$\langle p, s\rangle = \int\limits_Q p^T(q)B(q)s(q)dq, \quad p, s \in H_B, \tag{7.6}$$

where $B(q)$ is a positively determined symmetric matrix, $l \times l$, $Q = D \times [0,T]$. Further, if $B(q)$ is the unit matrix the index of the scalar product will be omitted.

Let a vector function $d(q)$ have a random character, be of a multidimensional normal distribution with a known mean-value $d_0(q)$ and a covariance matrix $\Gamma_d(q,q')$. Then, using the terminology of statistical interpretation theory, the inverse problem will be to find the optimal estimate $\hat{d}(q)$ of the vector function $d(q)$ using information (1.111).

Accordingly to theory (Gol'tsman, 1982), estimates by the method of *a posteriori* probability density $P(\hat{d})$ are asymptotically optimal, i.e., probability distribution is normal; mean estimates coincide with proper parameter values and the estimates have minimal dispersion. According to this method, as the estimate $\hat{d}$ we take the vector function d which gives the maximum of the *a posteriori* probability density $P(d|v)$

$$P(\hat{d}|v) = \sup_{d} P(d|v),$$

which, by the Bayes formula, is (Korlyuch, 1985)

$$P(d|v) = \text{const} P(d) P(v|d). \tag{7.7}$$

Here $P(d)$ is the *a priori* probability density; $P(v|d)$ is the conditional probability density. These values, by the given assumptions, are as follows

$$P(d) = \frac{\exp\left\{-\frac{1}{2}\left[\left(d(q') - d_0(q')\right)^T \Gamma_d^{-1}(q,q')\left(d(q) - d_0(q)\right)\right]\right\}}{(2\pi)^{1/2}\left(\det\left(\Gamma_d(q,q')\right)\right)^{1/2}}, \tag{7.8}$$

$$P(v|d) = \frac{\exp\left\{-\frac{1}{2}\left[\left(v(q') - u(q',d)\right)^T \Gamma_\varepsilon^{-1}(q,q')\delta_R\left(v(q) - u(q,d)\right)\right]\right\}}{(2\pi)^{1/2}\left(\det\left(\Gamma_\varepsilon(q,q')\delta_R\right)\right)^{1/2}}. \tag{7.9}$$

The index T in these formulas denotes transposition; δ_R is the generalized function of zero order, the carrier of which coincides with the manifold $R \subset D; \Gamma_d^{-1}, \Gamma_\varepsilon^{-1}$ are inverse matrices to the known covariance matricies Γ_d and Γ_ε respectively.

Lemma 1.7. Let covariance matrices Γ_d and Γ_ε be as follows

$$\Gamma_\delta(q,q') = G_d\delta(q-q') = \left(G_c(x)\delta(t) + G_f(q)\right)\delta(q-q'), \tag{7.10}$$

$$\Gamma_\varepsilon(q,q') = G_\varepsilon(q)\delta(q-q'), \quad q,q' \in R, \tag{7.11}$$

where G_d, G_ε are symmetric positively determined matrices $(L+N) \times (L+N)$ and $N \times N$ respectively. Then the problem of determining the optimal estimate $\hat{d}$ by the method of maximum *a posteriori* probability density is equivalent to determining the vector function d_* minimizing the functional

$$J(d) = \Big\langle \big(v - u(q;d)\big), \big(v - u(q;d)\big) \Big\rangle_{B_1} + \Big\langle \big(d - d_0\big), \big(d - d_0\big) \Big\rangle_{B_2} \qquad (7.12)$$

where

$$B_1(q) = G_\varepsilon^{-1}(q)\delta_R, \qquad (7.13)$$

$$B_2(q) = G_d^{-1}(q). \qquad (7.14)$$

Proof. As probabilities (1.114) and (1.115) depend on q, q', and the logarithm is monotone, the integral maximum of $P(d|v)$ coincides with the minimum of the function

$$\Lambda(d) = -\int_Q\int_Q \ln\big(P(d)P(v|d)\big)\,dq\,dq', \qquad (7.15)$$

which is called the "sufficient receiving function" (Vladimirov, 1976). Substituting in this expression probability densities (1.114), (1.115); noting the covariance function form (1.114), (1.115) and neglecting the terms and factors which have no influence on the minimum in the set of desired parameters $d(q)$, we obtain the functional (1.118). The latter has two terms. The first term takes into account the residual between the observed and model data. The second term takes into account *a priori* data on the desired vector d statistics. Thus, inverse problem (1.107)-(1.111) is formulated as the problem of determining the vector function d_* which gives the minimum to functional (1.118)

$$J(d_*) = \inf_d J(d).$$

Fréchet derivative of goal functional on the basis of the conjugate equation method and the perturbation algorithm. Parallel with setting (1.107)-(1.111), consider the conjugate initial-boundary-value problem. For example, in the case of hyperbolic operator $\mathcal{L}$

$$\mathcal{L}^* u^* = q, \tag{7.16}$$

$$Bu^* = 0, \tag{7.17}$$

$$u^*\Big|_{t=T} = 0, \tag{7.18}$$

$$u_t^*\Big|_{t=T} = 0. \tag{7.19}$$

The structure of the fundamental solution of the problem in the simplest case is given by (1.104). As was noted in remark in §1.6, the wave process continues for inverse flow of time (from T to 0). Evidently, for these settings in the case when $\varphi = 0$, the lemma conditions (from §1.6) and equality (1.101): $\langle f, u^* \rangle = \langle g, u \rangle$ hold. Here $\langle\ ,\ \rangle$ denotes the scalar product of vector function (1.112), where the weight matrix B is the unit matrix $N \times N$. First, consider the problems of existence and smoothness of the Fréchet derivative of functional (1.118).

Lemma 1.8. Let components of the vector function $d_0(q) = \big(f_0(x,t), c_0(x)\big)$ and elements of the covariance matrices G_f, G_c, G_ε be continuous in x and t. Assume this holds for direct problem (1.107)-(1.110) and conjugate problem (1.122)-(1.125) with the right side $g\big(v - u(q;d)\big) \cdot G_\varepsilon^{-1} \delta_R$ are unique in a certain class of sufficiently smooth functions; and for these functions the inequality

$$\int\limits_Q \left\| \frac{\partial u}{\partial d}^T \frac{\partial u}{\partial d} G_\varepsilon^{-1}(q) + G_d^{-1}(q) \right\|^2 dq \le \mu_0^2 \tag{7.20}$$

holds, where the norm $\|\ \cdot\ \|$ is the Euclidean norm of the matrix $(N+L) \times (N+L)$. Let, also,

$$\int_T \left(\frac{\partial t}{\partial c}\right)^T uu^* dt \in C(D).$$

Then functional (1.117) is differentiable by Fréchet and for $d \in C(D)$ has the continuous derivative

$$J'_d = 2\left(\int_T \left(\frac{\partial t}{\partial c}\right)^T uu^* dt + G_c^{-1}(c - c_0) + u^* + G_f^{-1}(f - f_0)\right). \quad (7.21)$$

Proof. We shall look for the derivative of functional (1.117) by the perturbance method (Korlyuch, 1985).

Let $u(d)$ solve (1.107) for some known unperturbed parameters, and $u(d + \delta d)$ be the solution for perturbation of smooth coefficients. Then the solution perturbation is $\delta u = u(d + \delta d) - u(d)$ and

$$J(d + \delta d) = \Big\langle \big(v - u - \delta u\big), \big(v - u - \delta u\big) \Big\rangle_{B_1}$$
$$+ \Big\langle \big(d + \delta d - d_0\big), \big(d + \delta d - d_0\big) \Big\rangle_{B_2}.$$

Hence

$$\delta J = J(d + \delta d) - J(d) = -2\Big\langle \big(v - u(d)\big), \delta u \Big\rangle_{B_1}$$
$$+ \langle d - d_0, \delta d \rangle_{B_2} + \langle \delta u, \delta u \rangle_{B_1} + \langle \delta d, \delta d \rangle_{B_2}. \quad (7.22)$$

The first two terms in the right side of this expression are linear relative to δd and describe the Fréchet derivative when the two latter terms are terms of the order $O(\langle \delta d, \delta d \rangle)$ Represent the solution variation as $\delta u = \partial u / \partial d \cdot \delta d$, and, using the Cauchy-Bunyakovskii inequality and estimate (1.126), we obtain

$$\langle \delta u, \delta u \rangle_{B_1} + \langle \delta d, \delta d \rangle_{B_2} = \int_Q \left[\left(\frac{\partial u}{\partial d}\delta d\right)^T G_\varepsilon^{-1} \left(\frac{\partial u}{\partial d}\delta d\right) + \delta d^T G_d^{-1} \delta d \right] dq$$

$$= \int_Q \left[\left(\frac{\partial u}{\partial d} G_\varepsilon^{-1} \frac{\partial u}{\partial d} \delta d \right)^T \delta d + \left(G_d^{-1} \delta d \right)^T \delta d \right] dq \le \mu_0^2 \langle \delta d, \delta d \rangle.$$

Thus, under condition (1.126), the Fréchet derivative exists. To find it, express the solution variation δu in (1.128) by means of parameter variation $\delta d = (\delta c, \delta f)$ which, in its turn, is connected in linear approximation with variation of the operator $\mathcal{L} : \delta \mathcal{L} = (\delta \mathcal{L}/\partial c)\delta c$. Here $\delta \mathcal{L}/\partial c$ is the matrix $N \times N$ with elements-columns of the length L.

On the basis of the perturbation algorithm for equation (1.107) we have

$$(\mathcal{L} + \varepsilon \delta \mathcal{L})(u + \varepsilon \delta u + \ldots) = f + \varepsilon \delta f,$$

where ε is a small parameter. Equating in both sides of this equality the corresponding coefficients, we obtain equation (1.107) and the equation

$$\mathcal{L} \delta u = -\delta \mathcal{L} u + \delta f. \tag{7.23}$$

Consider the equation $\mathcal{L}^* u^* = (v - u) \cdot G_\varepsilon^{-1} \delta_R$ conjugate to $\mathcal{L}$. Multiplying it scalarly by δu, equation (1.129) by u^*, and subtracting one from the other, we obtain (taking account of (1.101))

$$\left\langle \left((v - u) G_\varepsilon^{-1} \delta u \right), \delta u \right\rangle = \left\langle u^*, \left(-\delta \mathcal{L} u + \delta f \right) \right\rangle.$$

Therefore, the linear term of (1.128) by δd is as follows

$$\delta J = -2 \left\langle u^*, \left(\left(-\frac{\partial \mathcal{L}}{\partial c} \right)^T u \right)^T \delta c + \delta f \right\rangle + 2 \left\langle (d - d_0) G_d^{-1} \delta d \right\rangle$$

$$= \left\langle u^*, \left(\left(\frac{\partial \mathcal{L}}{\partial c} \right)^T u \right)^T \delta c \right\rangle - 2 \left\langle (u^*, \delta f) \right\rangle + 2 \left\langle (c - c_0) G_c^{-1} \delta c \right\rangle + 2 \left\langle (f - f_0) G_f^{-1} \delta f \right\rangle.$$

Hence the derivative column vector ($L \times N$) is given by (1.127). The first L components are determined by the first two terms in (1.127) and describe the functional derivative with respect to the coefficients $c(x)$ of operator $\mathcal{L}$. The remaining N components are determined by the second group of terms in (1.127) and describe the functional derivative with respect to the components of the source $f(x,t)$.

Due to the assumption on smoothness of the right side coefficients of (1.127), mean values of the covariance matrices components and condition (1.126), the components of the vector J'_d are continuous with respect to x and t.

Solvability of the inverse problem. Consider the problems of existence and uniqueness of the solution to the inverse problem of the right side and coefficients of equation (1.107).

On the basis of lemmas 1.7 and 1.8, we may prove the following theorem.

Theorem 1.9. Let $\mathcal{M}$ be the set of vector functions $d(q)$ such that

$$\Big\langle (d-d_0),(d-d_0)\Big\rangle_{B_2} \le \mu_1^2, \tag{7.24}$$

where the weight scalar product is given by (1.112), (1.120). Assume that the lemmas 1.8, 1.7 conditions hold and for each $d_1, d_2 \in \mathcal{M}$

$$\iint\limits_{D\,0}^{T}\left[u_1^*\left(\left(\frac{\partial\mathcal{L}}{\partial c_1}\right)^T u_1\right)^T - u_2^*\left(\left(\frac{\partial\mathcal{L}}{\partial c_2}\right)^T u_2\right)^T\right] dt(c_1(x)-c_2(x))dx + \mu_1^2$$

$$\ge \iint\limits_{D\,0}^{T} dx dt\,\Big((u_1^*)^T - (u_2^*)^T\Big)\Big(f_1(x,t)-f_2(x,t)\Big). \tag{7.25}$$

Then solution to inverse problem (1.107)-(1.114) exists, is unique in $\mathcal{M}$ and each minimizing sequence $\{d_k\} \in \mathcal{M}$, $k = 1,2,\dots$ weakly converges to this solution independent of initial approximation.

Proof. The set of correctness $\mathcal{M}$ is a convex subset of the weight Hilbert space $H_{B_2} \subset H$ with scalar product (1.115), (1.120). Really, for each $d_1, d_2 \in \mathcal{M}$ and $\alpha \in [0,1]$, the element $d_\alpha = \alpha d_1 + (1-\alpha)d_2$ belongs to $\mathcal{M}$, since, by the triangle inequality

$$||d_\alpha - d_0|| \leq ||d_\alpha - d_0||_{B_2} = ||\alpha d_1 + (1-\alpha)d_2 - d_0||_{B_2}$$

$$= ||\alpha(d_1 - d_0) + (\alpha - 1)d_0 + (1-\alpha)(d_2 - d_0)||_{B_2}$$

$$\leq ||d_1 - d_0||_{B_2} + (1-\alpha)||d_2 - d_0||_{B_2} \leq \alpha\mu_1 + (1-\alpha)\mu_1 = \mu_1.$$

Besides, the set $\mathcal{M}$ is weakly compact, since it is bounded (Butkovskii, 1979). Then, by the continuity, the functional $J(\alpha)$ (lemma 1.8) attains in $\mathcal{M}$ its greatest lower bound, i.e., the set

$$\mathcal{M}_* = \left\{ d_* | J(d_*) = \inf_{d \in \mathcal{M}} J(d) \right\}$$

is not empty.

To establish uniqueness, it is sufficient to show that the functional determined by the first term in (1.117)

$$J_1(d) = \left\langle \left(v - u(d)\right), \left(v - u(d)\right) \right\rangle_{B_1}$$

is convex. This is so, because the second functional (determined by the second term) is strongly convex, (which is easy to show); therefore, their sum will also be strongly convex. Hence follows that $\mathcal{M}_*$ has no more than one point (Vasiljev, 1981).

To establish that $J_1(d)$ is convex, use the known criterion: a functional is convex in a convex subset $\mathcal{M}$ of a Hilbert space H and only if

$$\left\langle \left(J'_d(d_1) - J'_d(d_2)\right), \left(d_1 - d_2\right) \right\rangle \geq 0 \;\; \forall d_1, d_2 \in \mathcal{M}.$$

Substituting in this inequality the functional derivatives from (1.127) and taking into account condition (1.129), which holds in $\mathcal{M}$, we obtain the desired condition (1.114). The theorem is proved.

§1.8 Formulas in multidimensional inverse problems for evolutionary equations.

Let $\mathcal{D}$ be a domain of the real Euclidean space R^{n+m+1} of the variables (x, z, t), $x = (x_1, \dots, x_m)$, $m \geq 0$, $t \in R$, $z = (z_1, \dots, z_n)$, $n \geq 1$.

In the domain $\mathcal{D}$, consider the equation

$$\frac{\partial u}{\partial t} = Au + \lambda(x,t)Bu + Lu. \tag{8.1}$$

Here $u(x, t, z)$, $\lambda(x, t)$, $(\partial\lambda/\partial z_k = 0, \ k = 1, 2, \dots, n)$ are the desired complex-valued functions; A, B, L are linear differential operators such that A and B act only on x and do not depend on (t, z). The operator L acts only on z and does not depend on (x, t).

Here are some examples of such an equation.

$$\frac{\partial u}{\partial t} = \sum_{j=1}^{m} \frac{\partial^2 u}{\partial x_j^2} + \frac{\partial^2 u}{\partial z^2} + \lambda(x,t)u, \quad A = \sum_{j=1}^{m} \frac{\partial^2}{\partial x_j^2}, \quad L = \frac{\partial^2}{\partial z^2}, \quad B = E;$$

$$\frac{\partial u}{\partial t} = -\frac{i}{n}\left(\sum_{j=1}^{m} \frac{\partial^2 u}{\partial x_j^2} + \frac{\partial^2 u}{\partial z^2}\right) + \lambda(x,t)u, \ A = -\frac{i}{n}\sum_{j=1}^{m} \frac{\partial^2}{\partial x_j^2},$$

$$L = -\frac{i}{n}\frac{\partial^2}{\partial z^2}; \quad B = E, \quad i^2 = -1;$$

$$A = \sum_{j=1}^{m} \frac{\partial^2 u}{\partial x_j^2} + \sum_{j=1}^{n} \frac{\partial^2 u}{\partial z_j^2} + \varphi(z,t)\lambda(x,t);$$

$$\frac{\partial u}{\partial t} = \sum_{j=1}^{m} \frac{\partial^2}{\partial x_j^2} + \sum_{j=1}^{n} \frac{\partial^2}{\partial z_j^2}; \quad Bu = \varphi(z,t);$$

where $\varphi(z, t)$ is a certain function.

When the functions $u(x, t, z)$, $\lambda(x, t)$ are to be determined simultaneously, the following initial-boundary information in usually given

$$u\Big|_{t=0} = u_0(x,z), \qquad u\Big|_{z=0} = f(x,t). \tag{8.2}$$

Below we consider the complex-valued functions on which the corresponding operators A, B, L are determined correctly.

Lemma 1.9. Let $f(x,y)$ be an arbitrary function such that $Bf \neq 0$ and let $\varphi(t,z)$ solve the equation $\frac{\partial\varphi}{\partial t} = L\varphi$ and $\varphi_0(t) = \varphi(t,0) \neq 0$. Then the functions

$$u(x,t,z) = \frac{f(x,t)\varphi(t,z)}{\varphi_0(t)}, \qquad \varphi_0(t) = \varphi\Big|_{z=0},$$

$$\lambda(x,t) = \left(\left(\frac{\partial f}{\partial t} - Af\right)\varphi_0 - f\frac{\partial\varphi_0}{\partial t}\right)\left(\varphi_0 Bf\right)^{-1},$$

satisfy equation (1.132) and the data

$$u\Big|_{z=0} = f(x,t), \quad u\Big|_{t=0} = \frac{f(x,0)\varphi(0,z)}{\varphi_0(0)} = u_0(x,z).$$

Let L be a differential operator with constant coefficients; let $c(\omega)$ be its symbol

$$L\left(e^{i\omega z}\right) = c(\omega)e^{i\omega z}, \quad \omega \in R^n, \quad z \in R^n.$$

Let $L = a^2 \sum\limits_{j=1}^{n} \frac{\partial^2}{\partial z_j^2}$, and, as above,

$$u\Big|_{t=0} = u_0(x,z) = g(x)\psi(z), \; u\Big|_{z=0} = f(x,t), \; g(x) = f(x,0), \; \psi(0) = 1.$$

If $\psi(z)$ is a bounded continuous function of $z \in R^n$ and the conditions of lemma 1.9 hold, then

$$\varphi(z,t) = \frac{1}{2a(\pi t)^{n/2}} \int\limits_{R^n} \psi(\zeta)\exp\left(-\frac{(\zeta - z)^2}{4ta^2}\right) d\zeta,$$

$$u(x,z,t) = \frac{f(x,t)\varphi(z,t)}{\varphi_0(t)}, \qquad \varphi_0(t) = \varphi(0,t),$$

$$\lambda(x,t) = \left(\left(\frac{\partial f}{\partial t} - Af\right)\varphi_0 - f\frac{\partial\varphi_0}{\partial t}\right)\left(\varphi_0(t)Bf\right)^{-1}.$$

Consider the problem of simultaneously determining two functions $u(x,z,t)$, $\lambda(x,t)$, $x \in \tilde{\mathcal{D}} \in R^n$ $z \in R^n$, $t > 0$, if, together with equation (1.132) we have conditions (1.133)

$$u\Big|_{t=0} = u_0(x,z), \qquad u\Big|_{z=0} = f(x,t).$$

Let L be an operator with the symbol $c(\omega)$, $\omega \in R^n$. If $u_0(x,z) = g(x)\psi(z)$, $g(x) = f(x,0)$, $\psi(0) = 1$, the conditions of lemma 1.9 hold and the Fourier transformation of $\psi(z)$ exists.

Then the following formulas hold.

$$a(\omega) = \int_{R^n} \psi(z)e^{-iz\omega}dz,$$

$$u(x,z,t) = \left(f(x,t)\int_{R^n} a(\omega)e^{i\omega z + c(\omega)t}d\omega\right)\left(\int_{R^n} a(\omega)e^{c(\omega)t}d\omega\right)^{-1};$$

$$\lambda(x,t) = \left(\left(\frac{\partial f}{\partial t} - Af\right)\int_{R^n} e^{c(\omega)t}a(\omega)d\omega\right.$$

$$\left.-f\int_{R^n} e^{c(\omega)t}ac\, d\omega\right)\left(Bf\int_{R^n} e^{c(\omega)t}a(\omega)d\omega\right)^{-1}.$$

Let $L = \sum\limits_{k=1}^{n}\frac{\partial^2}{\partial z_k^2}$, $u|_{z=0} = f(x,t)$, $u|_{t=0} = \varphi(x)\psi(z)$, $\varphi(x) = f(x,0)$, $\psi(z)$ be a continuous bounded function in R^n. Then the following formulas hold.

$$u(x,z,t) = \left(f(x,t)\int_{R^n}\psi(\zeta)exp\left(-\frac{|\zeta - z|^2}{4t}\right)d\zeta\right)\left(\int_{R^n}\psi(\zeta)exp\left(\frac{-\zeta^2}{4t}\right)d\zeta\right)^{-1},$$

$$\lambda(x,t) = \left(\left(\frac{\partial f}{\partial t} - Af\right)\frac{1}{2(\pi t)^{n/2}}\int_{R^n}\psi(\zeta)exp\left(-\frac{\zeta^2}{4t}\right)d\zeta\right.$$

$$\left.-f\frac{d}{dt}\left(\frac{1}{2(\pi t)^{n/2}}\right)\int_{R^n}\psi(\zeta)exp\left(-\frac{\zeta^2}{4t}\right)d\zeta\right)\left(Bf\cdot\int_{R^n}\psi(\zeta)exp\left(-\frac{\zeta^2}{4t}\right)d\zeta\right)^{-1}.$$

Lemma 1.9 yields

Theorem 1.10. Let (1.132), (1.133) hold, $L = \sum_{k=1}^{n} \frac{\partial^2}{\partial z_k^2}$, $u_0(x,z) = a(x)b(z)$, $\varphi(x) = f(x,0)$, $a(z), b(z)$ be continuous and bounded in R^n, $b(0) = 1$, and the conditions of lemma 1.9 hold. Then inverse problem (1.132), (1.133) is solved as follows

$$u(x,z,t) = \left(f(x,t)\int\limits_{R^n} b(\zeta) exp\left(-\frac{|\zeta - z|^2}{4t}\right) d\zeta\right)\left(\int\limits_{R^n} b(\zeta) exp\left(\frac{-\zeta^2}{4t}\right) d\zeta\right)^{-1},$$

$$\lambda(x,t) = \left(\left(\frac{\partial f}{\partial t} - Af\right)\frac{1}{2(\pi t)^{n/2}}\int\limits_{R^n} b(\zeta) exp\left(-\frac{\zeta^2}{4t}\right) d\zeta\right.$$

$$\left. - f\frac{d}{dt}\left(\frac{1}{2(\pi t)^{n/2}}\right)\int\limits_{R^n} b(\zeta) exp\left(-\frac{\zeta^2}{4t}\right) d\zeta\right)\left(\frac{Bf}{2(\pi t)^{n/2}}\cdot\int\limits_{R^n} b(\zeta) exp\left(-\frac{\zeta^2}{4t}\right) d\zeta\right)^{-1}.$$

Corollary. Let $Bu = \varphi(z,t)$ and the conditions of theorem 1.9 hold. Then the source function $h(x,z,t) = \varphi(z,t)\cdot\lambda(x,t)$ and the solution $u(x,z,t)$ to equation (1.132) is found from the information (1.133) as follows

$$f(z,t) = \frac{1}{(\pi t)^{1/2}}\int\limits_{R^n} b(\zeta) exp\left(-\frac{|\zeta - z|^2}{4t}\right) d\zeta,$$

$$u(x,z,t) = \frac{f(x,t)\varphi(z,t)}{\varphi_0(t)},$$

$$\lambda(x,t) = \left(\left(\frac{\partial f}{\partial t} - Af\right)\varphi_0 - f\frac{\partial \varphi_0}{\partial t}\right)\varphi\varphi_0^{-2},$$

$$\varphi_0(t) = \varphi(0,t).$$

Lemma 1.10. Let $f(x,y)$, $a(\omega)$, $(x,y) \in \mathcal{D}$, $\omega = R^n$ be complex-valued functions such that the expressions

$$\frac{\partial f}{\partial t} - Af, \quad Bf, \quad \int\limits_{R^n} e^{c(\omega)y + i\omega z}\cdot a(\omega) d\omega$$

are valid and

$$\int\limits_{R^n} e^{c(\omega)y} \cdot a(\omega) d\omega \neq 0, \qquad Bf \neq 0.$$

Then the functions

$$u(x,z,t) = \left(f(x,y) \int\limits_{R^n} a(\omega) e^{i\omega z + c(\omega)y} dy \right) \left(\int\limits_{R^n} a(\omega) e^{c(\omega)y} d\omega \right)^{-1};$$

$$\lambda(x,t) = \left(\left(\frac{\partial f}{\partial t} - Af \right) \int\limits_{R^n} e^{c(\omega)y} a(\omega) d\omega \right.$$

$$\left. - f \int\limits_{R^n} e^{c(\omega)y} c(\omega) a(\omega) d\omega \right) \left(Bf \int\limits_{R^n} e^{c(\omega)y} a(\omega) d\omega \right).$$

satisfy equation (1.132) and the data

$$u\Big|_{z=0} = f(x,y),$$

$$u\Big|_{y=0} = \varphi(x)\psi(z),$$

where

$$\varphi(x) = f(x,0);$$

$$\psi(z) = \left(\int\limits_{R^n} a(\omega) e^{i\omega z} d\omega \right) \left(\int\limits_{R^n} a(\omega) d\omega \right)^{-1}.$$

Remark. In lemma 1.9 as the function $\varphi(z,t)$ we may take a function which satisfies the equation

$$\frac{\partial \varphi}{\partial t} = L\varphi + K\varphi\varphi_0,$$

where K is a certain constant. In this case

$$\lambda(x,t) = \left(\left(\frac{\partial f}{\partial t} - Af \right) \varphi_0 - f \frac{\partial \varphi_0}{\partial t} + K\varphi_0^2 f \right) \left(\varphi_0^2 Bf \right)^{-1}.$$

The extension of functions $u(x, z, t)$ if it is needed may be done by adding to $u(x, z, t)$ the solution $R(x, z, t)$ to the equation

$$\frac{\partial R}{\partial t} = AR + \lambda BR + LR$$

with the condition

$$R\Big|_{z=0} = 0.$$

Chapter 2

Correctness of certain inverse problems.

In this chapter we consider different multidimensional inverse problems for the determination of sources, both in linear and in nonlinear cases. Here, we cite many results concerning the newly-posed problems, their solvability and uniqueness. More explicitly, these questions are investigated in Bubnov (1986, 1987a, b, c, d, 1988), Anikonov and Bubnov (1988a, b).

§2.1 The Cauchy problem for elliptic and ultrahyperbolic equations in the classes of finite smoothness functions and inverse problems.

When we solve inverse problems for hyperbolic and parabolic equations, we often meet the Cauchy problem for elliptic and other types of equations. The Cauchy problem in the class of analytic functions has been investigated quite thoroughly (Lavrent'ev *et al.*,1980), but in applications the problem arises of its solvability in classes of finite smoothness. Note that, as follows from the well-known Petrovskii theorem (Petrovskii, 1938), the Cauchy problem is ill-posed for elliptic equations in classes of functions of finite smoothness.

In this section we shall give the necessary and sufficient conditions of solvability for elliptic and ultrahyperbolic equations in classes of functions of finite smoothness, and we shall show how these conditions are effectively applied for inverse problem research.

Necessary and sufficient solvability conditions for the Cauchy problem for the nonlinear evolutionary equation.

Setting 1. In the domain $R_+^{p+q+1} = \{(x, y, t),\ t > 0,\ x \in R^p,\ y \in R^q\}$ consider the equation

$$Lu = u_{tt} + \sum_{|\alpha|=2m} a_\alpha D_x^\alpha u - (-1)^n \sum_{|\beta|=2n} b_\beta D_y^\beta u$$
$$+ a(x, y, t)u + a_0 u + \varepsilon g(t)u^2 = f(x, y, t), \tag{1.1}$$

where $\varepsilon > 0$, α_0, a_α, b_β are real constants; a, g are real-valued functions

$$\sum_{|\alpha|=2m} a_\alpha \xi^\alpha \geq c_0 |\xi|^{2m}, \quad \sum_{|\beta|=2n} b_\beta \eta^\beta \geq c_0 |\eta|^{2n}, \quad \xi \in R^p, \quad \eta \in R^q.$$

Note that the class of equations (2.1) contains for $m = n = q = p = 1$ the hyperbolic equations, and for $m = n = 1$, $p > 1$, $q < 1$ the ultrahyperbolic equations.

The Cauchy problem. To find in R_+^{p+q+1} the solution to (2.1) satisfying the conditions

$$u_t \Big|_{t=0} = u_1(x, y), \qquad u \Big|_{t=0} = u_0(x, y). \tag{1.2}$$

Denote $c(\xi, \eta) = \sum\limits_{|\alpha|=2m} a_\alpha \xi^\alpha - \sum\limits_{|\beta|=2n} b_\beta \eta^\beta = \alpha_0$, $\xi \in R^p$, $\eta \in R^q$ and fix $\delta > 0$ so that the sets

$$\mathcal{D}_1 = \Big\{(\xi, \eta) : |c(\xi, \eta)| \leq \delta^2\Big\}; \ \ \mathcal{D}_2 = \Big\{(\xi, \eta) : c(\xi, \eta) > \delta^2 > 0\Big\},$$
$$\mathcal{D}_3 = \Big\{(\xi, \eta) : c(\xi, \eta) < -\delta^2 < 0\Big\}$$

would be not empty.

Theorem 2.1. For each f, u_1, u_0, a, g such that

$$\int_0^\infty e^{-2\delta t}\left\|f\right\|^2_{W^{2m,2n}_{z,x,y}} dt < \infty,$$

$$\int_{R^{p+q}} \left\|e^{-\delta_1 t}\hat{f}\right\|_{L^\infty_{(0,\infty)}} \left(1+|\xi|^{2m}+|\eta|^{2n}\right) d\xi d\eta, \quad 0<\delta_1<\delta;$$

$$u_1 \in W^{2m,2n}_{z,x,y}, \quad \left(1+|\xi|^{2m}+|\eta|^{2n}\right)\hat{u}_1 \in L^1(R^{p+q});$$

$$u_0 \in W^{2m,2n}_{z,x,y}, \quad \hat{u}_0 : (\hat{u}_{01}, \hat{u}_{02}, \hat{u}_{03});$$

$$\left(1+|\xi|^{2m}+|\eta|^{2n}\right)\hat{u}_{0i} \in L^1(\mathcal{D}_i)\bigcap L^2(\mathcal{D}_i), \quad i=1,2, \quad \hat{u}_{0i} = \hat{u}_0\Big|_{\mathcal{D}_i};$$

where

$$\hat{u}_{03} = -\frac{\hat{u}_{13}}{\sqrt{-c}} - \frac{1}{\sqrt{-c}}\int_0^\infty \hat{f}_3 e^{-t\sqrt{-c}} dt + A, \quad (\xi,\eta)\in\mathcal{D}_3;$$

$$\left(1+|\xi|^{2m}+|\eta|^{2n}\right)A \in L^1(\mathcal{D}_i)\bigcap L^2(\mathcal{D}_i), \tag{1.3}$$

there exist ε_1, $\varepsilon_2 > 0$ such that if

$$\int_{R^{p+q}} \max_t \left|\hat{a}e^{(\lambda_0+\delta)t}\right|(t+1)^{1/2}(1+|\xi|^{2m}+|\eta|^{2n}) d\xi d\eta \le \varepsilon_1,$$

$$\xi \in [0,\xi_2], \quad \lambda_0 \in (0,1), \quad \max_t \left|g(t+1)e^{(2\delta+\lambda_0)t}\right| < \infty,$$

the necessary and sufficient condition of unique solvability of problem (2.1)-(2.2) from the class

$$u_{tt} \in L^\infty\left(0,T;\ L^2(R^{p+q})\right), \quad u \in L^\infty\left(0,T;\ W^{2m,2n}_{z,x,y}(R^{p+q})\right) \quad \forall T>0;$$

$$\left\|u\right\|^2_{L^2(R^{p+q})} = O\left((t+1)e^{2\delta t}\right); \quad \left\|\hat{u}\right\|_{L^1(R^{p+q})} = O\left((t+1)^{1/2}e^{\delta t}\right)$$

is condition (2.3), where

$$\hat{\psi}(t,\xi,\eta) = (2\pi)^{-(p+q)/2}\int_{R^{p+q}} \psi(t,x,y)e^{i(\xi,x)+i(\eta,y)} dx dy,$$

$$\hat{\psi}_k(t,\xi,\eta) = \hat{\psi}(t,\xi,\eta), \quad (\xi,\eta)\in\mathcal{D}_k, \quad k=1,2,3,$$

and the function A in (2.3) is uniquely determined by input data.

Proof of this theorem is based on solving the Cauchy problem

$$v''(t,\xi,\eta) + c(\xi,\eta)v(t,\xi,\eta) + \int\limits_{R^{p+q}} \hat{a}(\xi-\lambda,\eta-z,t)v(t,\lambda,z)d\lambda dz$$

$$+\varepsilon g^{(t)} \int\limits_{R^{p+q}} v(t,\xi-\lambda,\eta-z)v(t,\lambda,z)d\lambda dz = \Phi(t,\xi,\lambda),$$

$$v(0,\xi,\lambda) = v_0(\xi,\eta), \quad v'(0,\xi,\lambda) = v_1(\xi,\lambda)$$

in the class of functions

$$\int\limits_{R^{p+q}} \left|v(t,\xi,\lambda)\right|^2 \left(1+|\xi|^{2m}+|\eta|^{2n}\right)d\xi d\eta < \infty$$

Setting 2. In the domain $R^3_+ = \{(x,y,t): \ t>0, \ (x,y)\in R^2\}$ consider the equation

$$Lu \equiv u_{tt} + D_x^{2m}u + D_y^{2n}u + a_0(x,y,t)D_x^{2m}u + b_0(x,y,t)D_y^{2n}u$$

$$+\sum_{k=1}^{2m-1} a_k(x,y,t)D_x^k u + \sum_{k=1}^{2n-1} b_k(x,y,t)D_y^k u + \alpha(x,y,t)u$$

$$+\alpha_0 u + \varepsilon g(t)u^2 = f(x,y,t), \tag{1.4}$$

where $a_0, b_0, a_k, b_k, \alpha, g, f$ are real -valued functions; ε, α_0 are positive constants; n and m are odd.

The Cauchy Problem. To find in R^3_+ the solution to (2.4) such that

$$u_t\Big|_{t=0} = u_1(x,y), \quad u\Big|_{t=0} = u_0(x,y). \tag{1.5}$$

Denote $C(\xi,\eta) = \xi^{2m}+\eta^{2n}-\alpha_0$, fix $\delta>0$ and determine the sets

$$\mathcal{D}_1 = \left\{(\xi,\eta)\in R^2 : \left|c(\xi,\eta)\right| \le \delta^2\right\},$$

$$\mathcal{D}_2 = \left\{(\xi,\eta) \in R^2 : \left|c(\xi,\eta)\right| > \delta^2 > 0\right\}.$$

By $\psi_i(\xi,\eta,t)$ denote the contraction of $\psi(\xi,\eta,t)$ in the set $\mathcal{D}_i$, $i = 1,2$.

Let $\hat{a}_0, \hat{b}_0, \hat{a}_k, \hat{b}_k, \hat{\alpha} \in L^\infty(0,T;\ L^1(R^2))\,\forall T > 0$; then for the functions

$$v(\xi,\eta,t) \in L^\infty\left(0,T;\ L^1(R^2)\bigcap L^2(R^2)\right),$$

$$\left(1 + |\xi|^{2m} + |\eta|^{2n}\right)v \in L^\infty\left(0,T;\ L^1(R^2)\right),$$

define the convolution operator

$$\begin{aligned}
M(v_1,v_2) = &\int\limits_{R^2} \hat{u}_0(\xi-\lambda,\eta-z,t)v_1(\lambda,z,t)d\lambda dz \\
&+ \int\limits_{R^2} \hat{b}_0(\xi-\lambda,\eta-z,t)v_1(\lambda,z,t)d\lambda dz \\
&- \int\limits_{r^2} \hat{a}(\xi-\lambda,\eta-z,t)v_1(\lambda,z,t)d\lambda dz \\
&- \sum_{k=1}^{2m-1}(i)^k \int\limits_{R^2} \hat{u}_k(\xi-\lambda,\eta-z,t)\lambda^k v_1(\lambda,z,t)d\lambda dz \\
&- \sum_{k=1}^{2n-1}(i)^k \int\limits_{R^2} \hat{b}_k(\xi-\lambda,\eta-z,t)\eta^k v_1(\lambda,z,t)d\lambda dz \\
&- \varepsilon g(t)\int\limits_{R^2} v_1(\xi-\lambda,\eta-z,t)v_2(\lambda,z,t)d\lambda dz.
\end{aligned}$$

Consider the following Cauchy problem

$$\left.\begin{aligned}
&v''(\xi,\eta,t) - c(\xi,\eta)v - M(v,v) = \Phi(\xi,\eta,t),\\
&v(\xi,\eta,0) = v_0,\quad v'(\xi,\eta,0) = v_1.
\end{aligned}\right\} \tag{1.6}$$

Then in terms of v_1, v_2, problem (2.6) is equivalent to the following system

$$v_1'' - cv_1 - M(v,v) = \Phi_1,\ \ v_1(0) = v_{01},\ \ v_1' = v_{11}(\xi,\eta) \in \mathcal{D}_1,$$

$$v_1 = \frac{1}{2}\left(v_{02} - \frac{v_{12}}{\sqrt{c}}\right) e^{-t\sqrt{c}} - \frac{e^{-t\sqrt{c}}}{2\sqrt{c}} \int_0^t \left[\Phi_2 + M(v,v)\right] e^{\tau\sqrt{c}} d\tau$$
$$- \frac{e^{-t\sqrt{c}}}{2\sqrt{c}} \int_t^\infty \left[\Phi_2 + M(v,v)\right] e^{-\tau\sqrt{c}} d\tau$$
$$+ \frac{e^{t\sqrt{c}}}{2}\left[v_{02} + \frac{v_{12}}{\sqrt{c}} + \frac{1}{\sqrt{c}} \int_0^\infty \left(\Phi_2 + M(v,v)\right) e^{-t\sqrt{c}} d\tau\right],\ \ (\xi,\eta) \in \mathcal{D}_2. \tag{1.7}$$

Theorem 2.2 Let $\delta > \delta_1 > 0$, $\delta^2 > \alpha_0 \geq 0$. Then for each $f, u_1, u_0, a_0, b_0, a_k, b_k, \alpha, g$ such that

$$\int_0^\infty \int_{R^2} e^{-2\delta t} f^2 dx dy dt < \infty,\ \ \int_{R^2} \left\| e^{-\delta_1 t} \hat{f} \right\|_{L^\infty(0,\infty)} d\xi d\eta < \infty,$$
$$u_1 \in W^{m,n}_{2x,y}(R^2)(1 + \xi^{2m} + \eta^{2n})\hat{u}_1 \in L^1(R^2), \tag{1.8}$$

the function $u_0 \in W^{2m,2n}_{2x,y}(R^2)$, which is uniquely determined by its Fourier transform $\hat{u}_0 : (\hat{u}_{01}, \hat{u}_{02})$, is as follows

$$\hat{u}_{01} \in L^2(\mathcal{D}_1),$$

$$\hat{u}_{02} = -\frac{\hat{u}_{12}}{\sqrt{c}} - \frac{1}{\sqrt{c}} \int_0^\infty \hat{f}_2 e^{-t\sqrt{c}} dt + A,\ \ (\xi,\eta) > \delta^2 > 0,$$
$$\max(z_0, z_1) \leq \frac{1}{6},\ \ 0 \leq \xi \leq \xi_0. \tag{1.9}$$

Then the necessary and sufficient condition of unique solvability of problems (2.4), (2.5) in the class

$$u_{tt} \in L^\infty\left(0, T;\ L^2(R^2)\right),\ \ u \in L^\infty\left(0, T;\ W^{2m,2n}_{2x,y}(R^2)\right)\ \ \forall T > 0,$$

$$\left\|u\right\|^2_{L^2(R^2)} = O\left(e^{2\delta t}(t+1)\right), \ \left\|\hat{u}\right\|_{L^1(R^2)} = O\left(e^{\delta t}(t+1)^{1/2}\right),$$

$$(1+\xi^{2m}+\eta^{2n})\hat{u} \in L^1(R^2),$$

is condition (2.9). The function A in (2.9) is uniquely determined by input data; z_i are determined in (2.9).

Proof of theorem 2.2 is based on solvability of system (2.7) in the class of functions

$$\int_{\mathcal{D}_1}(1+\xi^{2m}+\eta^{2n})|v_1|^2 d\xi d\eta + \int_{\mathcal{D}_2}(1+\xi^{2m}+\eta^{2n})^2|v_2|^2 d\xi d\eta < \infty.$$

Example 1. Consider the Cauchy problem

$$u_{tt} + u_{xx} = f(x,t), \ \ t>0, \ \ x \in R,$$

$$u\Big|_{t=0} = u_0, \ \ u_t\Big|_{t=0} = u_1.$$

In this case the solvability conditions (2.9) will be as follows

$$u_1 \in W_2^1(R), \ \ \hat{u}_{01} \in L^2(\mathcal{D}_1),$$

$$u_{02} = \frac{\hat{u}_{12}}{|\xi|} - \frac{1}{|\xi|}\int_0^\infty \hat{f}_2 e^{-t|\xi|}dt, \ \ |\xi| > \delta > 0.$$

Example 2.

$$u_{tt} - u_{xx} - u_{yy} = f(t,y,x), \ \ t>0, \ \ (x,y) \in R^2,$$

$$u\Big|_{t=0} = u_0, \ \ u_t\Big|_{t=0} = u_1.$$

In this case solvability conditions are as follows

$$u_1 \in W_2^2(R^2), \ \ \hat{u}_0 : (\hat{u}_{01}, \hat{u}_{02}, \hat{u}_{03}),$$

$$\hat{u}_{0k}(1+\xi^2+\eta^2) \in L^1(\mathcal{D}_k) \bigcap L^2(\mathcal{D}_k), \ \ k=1,2,$$

$$\hat{u}_{03} = -\frac{\hat{u}_{13}}{\sqrt{\eta^2-\xi^2}} - \frac{1}{\sqrt{\eta^2-\xi^2}} \int\limits_0^\infty \hat{f}_3(t,\xi,\eta) e^{-t\sqrt{\eta^2-\xi^2}} dt,$$

$$\mathcal{D}_3 = \left\{ (\xi,\eta) : \eta^2 - \xi^2 > \delta^2 \right\},$$

where ξ, η are the dual variables to x, y.

Necessary and sufficient conditions for determining the right side of the wave equation. In the domain $Q = \{z > 0,$ $t > 0,\ x \in R^n\}$ consider the equation

$$u_{tt} - u_{zz} - \Delta_x u = h(t) f(x,z), \tag{1.10}$$

$$\Delta_x u = \sum_{t=1}^n \frac{\partial^2 u}{\partial x_i^2}.$$

Initial boundary-value problems. To find solution to (2.10) which satisfies the conditions

$$u \Big|_{t=0} = u_t \Big|_{t=0} = 0, \qquad u_z \Big|_{z=0} = u_1. \tag{1.11}$$

Inverse problems. To find functions u, f (h is a given function) entering in (2.10) such that (2.11) holds and

$$u \Big|_{z=0} = u_0 \tag{1.12}$$

from the class

$$e^{-p_0 t} u, \ \ e^{-p_0 t} D_t^\alpha u_{zz} \in L^\infty_{z,t,x} \left(0, \infty;\ L^2\left((0,\infty) \times R^n\right)\right), \ \ \alpha \le m-2;$$

$$e^{-p_0 t} D_x^\beta D_t^\gamma u \in L^\infty_{z,t,x}\Big(0,\infty; L^2\Big)\Big((0,\infty)\times R^n\Big), \ \ |\beta| < m, \ \ \gamma \le m;$$

$$f(x,0) = 0; \ \ f(x,z), \ \ D_x^\alpha D_z^\beta f \in L^\infty_{z,x}\Big(0,\infty; \ L^2(R^n)\Big),$$

$$|\alpha| \le m, \ \ \beta \le m, \ \ m \ge 2, \ \ p_0 > 0. \qquad (2.12^*)$$

We shall be interested in the case when *supp* $h(t) \subset (T_1, T_2)$, $0 < T_1 < T_2 < \infty$. The case when $h(0) \ne 0$ will be considered below.

To solve inverse problem (2.10)-(2.12) in the half-space $z > 0$, $x \in R^n$ consider the Cauchy problem for the elliptic equation

$$-v_{zz} - \Delta_x v + p^2 v = \hat{h}(p) f(x,z), \ \ z > 0, \qquad (1.13)$$

$$v\Big|_{z=0} = v_0, \qquad v_z\Big|_{z=0} = v_1, \qquad (1.14)$$

$$\hat{h}(p) = \int_0^\infty e^{-pt} h(t)dt; \qquad v_i = \int_0^\infty e^{-pt} u_i(x,t)dt, \ \ i = 0,1,$$

$$p = p_1 + ip_2, \ \ p_1 \ge p_0 > 0, \ \ \int_{-\infty}^{+\infty} (1+|p|)^{2m} \Big\|v_i\Big\|^2_{W_2^m} dp_2 < \infty,$$

$$h \in \overset{0}{W}{}_2^m(T_1, T_2).$$

As follows from the above results, problem (2.13), (2.14) for arbitrary v_i, h, f in the classes of finite smoothness functions is ill-posed. Find, as before, the necessary and sufficient conditions for initial data and the right side which guarantee well-posedness of this setting.

Consider the Cauchy problem for the ordinary differential equation

$$-\Phi_{zz} + (|\xi|^2 + p^2)\Phi = \hat{h}(p) \cdot \hat{f}(\xi, z), \ \ z > 0; \qquad (1.15)$$

$$\Phi(0) = \Phi_0(\xi, p), \quad \Phi'(0) = \Phi_1(\xi, p); \qquad (1.16)$$

$$\hat{f} = \frac{1}{(2\pi)^n} \int\limits_{R^n} f(x,z) e^{i(\xi,x)} dx, \quad \Phi_i = \frac{1}{(2\pi)^n} \int\limits_{R^n} v_i e^{i(\xi,x)} dx.$$

Denote

$$c(\xi,p) = |\xi|^2 + p^2, \quad \mathcal{D}_1 = \left\{(\xi,p) : |\xi|^2 + p_1^2 - p_2^2 \geq 0\right\};$$

$$\mathcal{D}_2 = \left\{(\xi,p) : |\xi|^2 + p_1^2 - p_2^2 < 0\right\}, \quad a = |\xi|^2 + p_1^2 - p_2^2;$$

$$\sqrt{c}\Big|_{\mathcal{D}_l} = \chi(p_2)\left[\frac{1}{2}\sqrt{a + \sqrt{a^2 + 4p_1^2 p_2^2}} + i\frac{\sqrt{2}p_1 p_2}{\sqrt{a + \sqrt{a^2 + 4p_1^2 p_2^2}}}\right];$$

$$l = 1,2; \qquad \chi(p_2) = \begin{cases} 1, & p_2 \geq 0, \\ -1, & p_2 < 0. \end{cases}$$

Lemma 2.1. The necessary and sufficient condition for solvability of (2.15), (2.16) from the class

$$I(\Phi) \equiv \int\limits_{-\infty}^{+\infty} |\Phi|^2 \int\limits_{R^n} \left(1 + |\xi|^2 + |p|^2\right)^m d\xi dp_2$$

$$+ \int\limits_{-\infty}^{+\infty} \int\limits_{R^n} |\Phi_{zz}|^2 \left(1 + |p|^2\right)^{m-2} d\xi dp_2 < \infty, \quad m \geq 2$$

is the condition

$$\Phi_0 + \frac{\Phi_1 \chi(p_2)}{\sqrt{c}} - \frac{\chi(p_2)}{\sqrt{c}} \int\limits_0^\infty \hat{h}(p) \hat{f}(\xi,z) e^{-\tau\sqrt{c}\chi(p_2)} d\tau = 0 \tag{1.17}$$

$$\forall \xi \in R^n, \quad p = p_1 + ip_2, \quad p_1 \geq p_0 > 0.$$

Besides, the following estimate

$$I(\Phi) \leq \text{const} \int\limits_{-\infty}^{+\infty} \int\limits_{R^n} \left(|\Phi_0|^2 + \frac{|\Phi_1|^2}{p_0^2}\right) \left(1 + |\xi|^2 + |p|^2\right)^m d\xi dp_2$$

$$+ \frac{\text{const}}{p_0^4} \left\| \int\limits_{-\infty}^{+\infty} \int\limits_{R^n} |\hat{h}(p)|^2 |\hat{f}|^2 \left(1 + |\xi|^2 + |p|^2\right)^m d\xi dp_2 \right\|_{L^\infty(0,\infty)}$$

holds.

Proof. Denote $\psi_j = \Phi\Big|_{\mathcal{D}_j}$, $j = 1, 2$; $\Phi = (\psi_1, \psi_2)$, $\Phi_0 = (\Phi_{01}, \Phi_{02})$, $\Phi_1 = (\Phi_{11}, \Phi_{12})$, $\Phi_{jk} = \Phi_j\Big|_{\mathcal{D}_k}$, $i = 0, 1$, $k = 1, 2$. Then the solution to (2.15), (2.16) may be written as follows

$$\psi_i = \frac{e^{\chi(p_2)z\sqrt{c}}}{2}\left[\Phi_{0i} + \frac{\chi(p_1)\Phi_{1i}}{\sqrt{c}} - \frac{\chi(p_2)}{\sqrt{c}}\int\limits_0^\infty \hat{h}(p)\hat{f}(\xi, z)e^{-\chi(p_2)\tau\sqrt{c}}d\tau\right]\Bigg|_{\mathcal{D}_i}$$

$$+\frac{e^{-\chi(p_2)z\sqrt{c}}}{2}\left(\Phi_{0i} + \frac{\chi(p_2)\Phi_{1i}}{\sqrt{c}}\right) + \frac{\chi(p_2)}{2\sqrt{c}}\int\limits_z^\infty e^{(z-\tau)\chi(p_2)\sqrt{c}}\hat{h}(p)\hat{f}(\xi, \tau)d\tau\Bigg|_{\mathcal{D}_i}$$

$$+\frac{\chi(p_2)}{2\sqrt{c}}\int\limits_0^z \hat{h}(p)\hat{f}(\xi, \tau)e^{-(z-\tau)\chi(p_2)\sqrt{c}}d\tau\Bigg|_{\mathcal{D}_i}, \quad i = 1, 2. \tag{1.18}$$

Note that $|\sqrt{c}| \geq p_0$, $|\mathrm{Re}\sqrt{c}|\Big|_{\mathcal{D}_i} \geq p_0$, $i = 1, 2$.

If (2.17) holds, then from (2.18) we obtain that $\Phi = (\psi_1, \psi_2)$ belongs to the class pointed out in lemma 2.1.

Let $\Phi = (\psi_1, \psi_2)$ belong to this class. Then (2.18) yields (2.17).

Lemma 2.2. For each v_0, v_1, $\hat{h}$, f such that

$$\int\limits_{-\infty}^{+\infty}\left(\left\|v_0\right\|^2_{W_2^m} + \left\|v_1\right\|^2_{W_2^m}\right)\left(1 + |p|^2\right)^m dp_2$$

$$+\left\|\,\left\|f\right\|^2_{W_2^m}\int\limits_{-\infty}^{+\infty}\left(1 + |p|^2\right)^m |\hat{h}|^2 dp_2\right\|_{L^\infty(0,\infty)} < \infty, \quad \mathrm{Re}p \geq p_0$$

the necessary and sufficient solvability condition for problem (2.13), (2.14) from the class

$$\int\limits_{-\infty}^{+\infty}\left(1 + |p|^2\right)^m \left\|v\right\|^2_{W_2^m(R^n)} dp_2 + \int\limits_{-\infty}^{+\infty}\left(1 + |p|^2\right)^{m-2}\left\|v_{zz}\right\|^2_{L_2^m(R^n)} dp_2 < \infty$$

is condition (2.7).

Proof. The proof follows from lemma 2.1 and the Parseval equality.

Consider condition (2.17)

$$\hat{h}(p)\int_0^\infty \hat{f}(\xi,z)e^{-\tau\chi(p_2)\sqrt{c}}d\tau = \sqrt{c}\left(\Phi_0 + \frac{\Phi_0}{\sqrt{c}}\chi(p_2)\right)\chi(p_2)\ . \tag{1.19}$$

Since $\hat{h}(p)$ is analytic in Re$p \geq p_0$, it may vanish only in isolated points. Therefore, (2.19) may be written as

$$\int_0^\infty \hat{f}(\xi,z)e^{-\tau\lambda}d\tau = F(\xi,\lambda), \qquad \lambda = \chi(p_2)\sqrt{c}\ , \tag{1.20}$$

where the function

$$F(\xi,\lambda) = \frac{\lambda}{\hat{h}\sqrt{\lambda^2-|\xi|^2}}\left[\Phi_0\left(\xi,-\sqrt{\lambda^2-|\xi|^2}\right)+\frac{\Phi_1}{\lambda}\left(\xi,\sqrt{\lambda^2-|\xi|^2}\right)\right] \tag{1.21}$$

is known in the domain $G = \{\lambda : \text{Re}\lambda \geq p_0,\ \text{Re}\sqrt{\lambda^2-|\xi|^2} \geq p_0\}$.

For each fixed ξ the domain G is the set $\{\lambda = \lambda_1 + i\lambda_2$, $\text{Re}\lambda \geq \sqrt{p_0^2+|\xi|^2}\}$ in the plane $\lambda = \lambda_1 + i\lambda_2$. (The branch $\sqrt{z}$ is taken so that $\text{Re}\sqrt{z} > 0$).

The function $F(\xi,\lambda)$, as follows from (2.20), is for fixed ξ the Laplace transform of $\hat{f}(\xi,\tau)$ by τ.

Theorem 2.3. Let the functions u_0, u_1, h satisfy the conditions.

a) The functions u_0, u_1 have the Laplace transform by t for Re$p \geq p_0 > 0$, $h \in \overset{0}{W}{}_2^m(T_1,T_2)$, $T_1 \geq 0$;

b)

$$\int_{-\infty}^{+\infty}\left(\|v_0\|^2_{W_2^m} + \|v_1\|^2_{W_2^m} + |\hat{h}|^2\right)\left(1+|p|^2\right)^m dp_2 < \infty, \quad m \geq 2,$$

where $v_0, v_1, \hat{h}$ are the Laplace transforms of u_0, u_1, h by t.

c) The functions

$$\Phi_0\left(\xi, \sqrt{\lambda^2 - |\xi|^2}\right), \quad \Phi_0\left(\xi, \sqrt{\lambda^2 - |\xi|^2}\right), \quad \hat{h}\left(\sqrt{\lambda^2 - |\xi|^2}\right)$$

are analytic by λ in the domain $\mathrm{Re}\lambda \geq \sqrt{\lambda^2 - |\xi|^2}$. Φ_0, Φ_1 are the Fourier transforms by x of the functions $v_0(x,p)$, $v_1(x,p)$; $\lambda = \sqrt{c(\xi, p)}$, $c = p^2 + |\xi|^2$, $\xi \in R^n$, $p = p_1 + ip_2$, $p_1 \geq p_0 > 0$.

Then the necessary and sufficient condition for unique solvability of inverse problem (2.10)-(2.12) from the class (2.12*) is the condition

$$\left\| \int_{R^n} \left(1 + |\xi|^2\right)^m \left[\left| \int_{-\infty}^{+\infty} F(\xi, \lambda) e^{\lambda z} d\lambda_2 \right|^2 \right.\right.$$

$$\left.\left. + \left| \int_{-\infty}^{+\infty} F(\xi, \lambda) e^{\lambda z} \lambda^m d\lambda_1 \right|^2 \right] d\xi \right\|_{L^\infty(0,\infty)} < \infty,$$

where the function F is determined by (2.21).

Proof of the theorem follows from lemma 2.2, the Parseval equality and the fact that from (2.20) follows that $\hat{f}$ is the inverse Laplace transform of the function F

$$\hat{f}(\xi, z) = \frac{1}{2\pi} \int_{-\infty}^{+\infty} F(\xi, \lambda) e^{\lambda z} d\lambda_2.$$

The inverse problem for the hyperbolic equation with singular initial data. In the domain $\{(t,x,z),\ x \in (0,A),\ |z| < \infty,\ t > 0\}$ consider the problem

$$u_{tt} - u_{xx} - u_{zz} + \alpha u_x + a(x,z)u = f(t,x,z)\theta(t-z), \tag{1.22}$$

$$D_x^k u\Big|_{x=0} = D_x^k u\Big|_{x=A}, \quad k = 0,1; \quad u\Big|_{t=0} = 0, \quad u_t\Big|_{t=0} = 2\delta(z)g(x), \tag{1.23}$$

$$g(x) = \sum_{|l|\geq 1} g_0 e^{i\eta_l x}; \quad f = \sum_{\substack{|l|\geq 1\\ |k|\geq 1}}^{n} e^{i\lambda_k t + i\eta_l x} f_{kl}(z); \quad \eta_l = \frac{2\pi l}{A}; \quad \lambda = \frac{2\pi k}{T};$$

$|\alpha| = \text{const} > 0$; $\theta(t)$ is the Heaviside function;

$$\frac{1}{g} \sum_m \bar{g}_m e^{i\eta_l x}; \qquad |g| \geq \delta > 0.$$

As in Romanov (1980), Romanov and Pukhnacheva (1987), we may show that (2.22), (2.23) is uniquely solvable and the solution is

$$u = \begin{cases} \tilde{u}, & t > |z|, \\ g(x), & t = |z|, \\ 0, & t < |z|, \end{cases}$$

where the smoothness of $\tilde{u}$ is determined by the smoothness of f, g, a.

Inverse problem. To find functions $u(t, x, z)$, $a(x, z)$ for $z > 0$. The function $a(x, z)$ is known for $z \leq a$ and $f|_{t=z} = f_0(x)$ from the class K of the finite smoothness functions entering in (2.22), satisfying (2.23) and

$$u_z\Big|_{z=0} = \theta(t) u_1(x, t) \equiv \sum_{\substack{|k|\geq 1\\ |l|\geq 1}}^{n} e^{i(\eta_l x + \lambda_k t)} u_{1kl} \theta(t). \tag{1.24}$$

Unlike the other settings, in this setting we give only one condition for $z = 0$ (not two conditions, as in Romanov and Pukhnacheva (1987)), for the problems of uniqueness, stability and local solvability in the classes of analytic by x, t functions. This fact arises because the problem is multidimensional and we consider it in the finite smoothness classes.

The class K, where the inverse problem is investigated, define as follows

$$K = \Big\{ u(t, x, z) : u = \theta(t - z) v(t, x, z), v\Big|_{t=x} = g(x),$$

$$v = \sum_{\substack{|k|\geq 1\\ |l|\geq 1}} e^{i(\eta_l x+\lambda_k t)} c_{kl}(z), c'_{kl}(z), c''_{kl}(z) \in L^\infty(0,\infty)\Big\}.$$

Consider for $z > 0$ the problem

$$-c''_{kl} + \gamma_{kl}c_{kl} + \sum_m h_m \tilde{c}_{k,-m+l} + 2\sum_m \bar{g}_m \sum_p \tilde{c}_{k,l}$$

$$\times \sum_{|n|\geq 1} e^{i\lambda_n z}(\lambda_n^2 \tilde{c}_{n,-m-p+l} - i\lambda_k c'_{n,-m-p+l}) = f_{k,l}, \qquad (1.25)$$

$$c'_{kl}(0) = u_{1kl}, \qquad |k| \geq 1, \qquad |l| \geq 1,$$

$$\tilde{c}_{kl} = \begin{cases} c_{kl}, & |l| \geq 1, \\ 0, & l = 0, \end{cases} \qquad \gamma_{kl} = -\lambda_k^2 + \eta_l^2 + i\alpha\eta_l.$$

Lemma 2.2. Let the functions c_{kl} solve problem (2.25). Then the functions

$$u = \theta(t-z)\sum_{\substack{|k|\geq 1\\ |l|\geq 1}} e^{i(\lambda_k t+\eta_l x)} c_{kl}(z),$$

$$a(x,z) = \frac{1}{g(x)}\left[f_0 + g_{xx} - \alpha g_x + 2\sum_{\substack{|l|\geq 1\\ |k|\geq 1}} (\lambda_k^2 c_{kl} - i\lambda_k c'_{kl}) e^{i(\lambda_k t+\eta_l x)}\right]$$

solve inverse problem (2.22)-(2.24) from K.

Remark. By solution to (2.25) are meant functions $c_{kl} : c_{kl}, c''_{kl} \in L^\infty(0,\infty)$ such that the series $\sum e^{i(\lambda_k t+\eta_l x)} c_{kl}(z)$ may be differentiated term by term and they converge in L^2.

Denote

$$\mathcal{D}_1 = \left\{(k,l) : -\lambda_k^2 + \eta_l^2 \geq 0\right\}, \quad \mathcal{D}_2 = \left\{(k,l) : -\lambda_k^2 + \eta_l^2 < 0\right\},$$

$$c^+_{kl}(z) = c_{kl}(z)\Big|_{\mathcal{D}_1}, \quad c^-_{kl}(z) = c_{kl}(z)\Big|_{\mathcal{D}_2}, \quad c_{kl}(z) = (c^+_{kl}, c^-_{kl}).$$

As before, it may be shown that solvability of (2.25) is equivalent to solvability of the system

$$c^{+}_{kl} = \frac{e^{-z\operatorname{sign}(æ)\sqrt{\gamma_{kl}}}}{2}\Big[-\frac{2\operatorname{sign}(æ)}{\sqrt{\gamma_{kl}}}u^{\pm}_{1kl}$$

$$-\frac{\operatorname{sign}(æ)}{\sqrt{\gamma_{kl}}}\int_0^\infty F e^{-\tau\operatorname{sign}(æ)\sqrt{\gamma_{kl}}}d\tau\Big]\Big|_{\mathcal{D}_i}$$

$$-\frac{\operatorname{sign}(æ)}{2\sqrt{\gamma_{kl}}}e^{z\operatorname{sign}(æ)\sqrt{\gamma_{kl}}}\cdot\int_z^\infty F e^{-\tau\operatorname{sign}(æ)\sqrt{\gamma_{kl}}}d\tau\Big|_{\mathcal{D}_i}$$

$$-\frac{\operatorname{sign}(æ)}{2\sqrt{\gamma_{kl}}}e^{-z\operatorname{sign}(æ)\sqrt{\gamma_{kl}}}\int_0^z F e^{\tau\operatorname{sign}(æ)\sqrt{\gamma_{kl}}}d\tau\Big|_{\mathcal{D}_i},$$

$$c^{\pm}_{kl} = -\frac{\operatorname{sign}(æ)}{2}e^{-z\operatorname{sign}(æ)\sqrt{\gamma_{kl}}}\Big[-2\operatorname{sign}(æ)u^{\pm}_{1kl}$$

$$-\operatorname{sign}(æ)\int_0^\infty F e^{-\tau\operatorname{sign}(æ)\sqrt{\gamma_{kl}}}d\tau\Big]\Big|_{\mathcal{D}_i}$$

$$-\frac{e^{-z\operatorname{sign}(æ)\sqrt{\gamma_{kl}}}}{2}\int_z^\infty F e^{-\tau\operatorname{sign}(æ)\sqrt{\gamma_{kl}}}d\tau\Big|_{\mathcal{D}_i}$$

$$+\frac{e^{-z\operatorname{sign}(æ)\sqrt{\gamma_{kl}}}}{2}\int_0^z F e^{\tau\operatorname{sign}(æ)\sqrt{\gamma_{kl}}}d\tau\Big|_{\mathcal{D}_i};$$

$$F = \sum_m h_m \tilde{c}_{k,-m+l} - f_{kl}$$

$$+2\sum_m \bar{g}_m \sum_p \tilde{c}_{kp} \sum_n \exp(i\lambda_n \bar{\bar{z}})(\lambda_n^2 \tilde{c}_{n,-m-p+l} - i\lambda_n \tilde{c}'_{n,-m-p+l}). \quad (1.26)$$

§2.2 Inverse problems for multidimensional hyperbolic equations.

Inverse problem for hyperbolic equations in the classes of functions that are periodical with respect to t. In the domain $Q_{2T} = (0, 2T) \times \mathcal{D}$, where $\mathcal{D}$ is a bounded domain in R^n with a boundary $S \in C^4$ (or $\mathcal{D} = \prod_{i=1}^{n} (0 < x_i < R)$), consider the equation

$$u_{tt} - \sum_{i,j=1}^{n} \Big(a_{ij}(x) u_{x_i}\Big)_{x_j} + a(x)u = F(x,t), \tag{2.1}$$

$$a_{ij} = a_{ji} \in C^4(\bar{\mathcal{D}}), \quad \sum_{i,j=1}^{n} a_{ij}\xi_i\xi_j \geq \mu|\xi|^2, \quad Au = \sum_{i,j=1}^{n} \Big(a_{ij}(x) u_{x_i}\Big)_{x_j}.$$

Boundary problem. To find in Q_{2T} a solution to (2.27) with conditions

$$D_t^k u\Big|_{t=0} = D_t^k u\Big|_{t=2T}, \quad k = 0, 1, \tag{2.2}$$

$$u\Big|_{\Gamma} = 0, \qquad \Gamma = (0, 2T) \times S. \tag{2.3}$$

Inverse problem 1. $a \in C^2(\bar{\mathcal{D}})$, $a \leq 0$. To find functions $u(x,t), h(x)$ ($F = h(x)f(x,t)$, f is given) such that (2.27)-(2.29) hold and

$$h(x) \in W_2^2(\mathcal{D}), \quad u = \sum_{k=0}^{\infty} \Big[u_k(x)\cos\lambda_k t + v_k(x)\sin\lambda_k t\Big], \quad \lambda_k = \frac{\pi k}{T};$$

$$u, D_t u \in L^2\Big(0, 2T; W_2^2(\mathcal{D})\Big), \quad u, D_t^3 u \in L^2(Q_{2T}), \quad u\Big|_{t=0} = \tau(x). \tag{2.4}$$

Inverse problem 2. $\mathcal{D} = \prod_{i=1}^{n} (0 < x_i < R)$, $n \leq 3$, $a_{ij} = \delta_{ij}$, $a = 0$. To find functions $u(x,t), h(x)$ ($F = h(x)f(x,t)$, f is given) in (2.27) such that (2.28), (2.29) hold and

$$h(x) \in W_2^2(\mathcal{D}), \quad u = \sum_{k=0}^{\infty} \Big[u_k(x) \cos \lambda_k t + v_k(x) \sin \lambda_k t \Big], \quad \lambda_k = \frac{\pi k}{T};$$

$$u, D_t^6 u \in L^2(Q_{2T}), \quad D_t^4 u \in L^2\Big(0, 2T; W_2^2(\mathcal{D})\Big), \quad u\Big|_{t=0} = \tau(x). \quad (2.5)$$

Remark 1. It should be noted that problem (2.27)-(2.29) is ill-posed. Its solvability depends on the geometry of $\mathcal{D}$ and on T.

We shall assume, in addition, that the following conditions hold

$$f, D_t^3 f \in L^2\Big(0, 2T; W_2^2(\mathcal{D})\Big), \quad f\Big|_{t=0} = \nu(x) \in W_2^2(\mathcal{D}) \bigcap C(\bar{\mathcal{D}}),$$

$$\min_{\bar{\mathcal{D}}} |\nu(x)| > 0, \quad f = \sum_{k=0}^{\infty} \Big[f_k(x) \cos \lambda_k t + g_k(x) \sin \lambda_k t \Big],$$

$$\tau(x) \in W_2^4(\mathcal{D}) \bigcap C(\bar{\mathcal{D}}), \quad \tau\Big|_S = 0. \quad (2.6)$$

For the inverse problem 2

$$f = \sum_{k=0}^{\infty} \left[f_k(x) \cos \lambda_k t + g_k(x) \sin \lambda_k t \right], \quad g_0 = 0, \quad \tau \in W_2^4(\mathcal{D}) \bigcap \overset{0}{W}{}_2^1(\mathcal{D}),$$

$$\pi^2 \left| \frac{k^2}{T^2} - R^{-2} \sum_{i=1}^{n} m_i^2 \right| \geq \frac{\pi^2}{R^2} \frac{C\left(R^2/T^2\right)}{K^{2d-2}}, \quad (2.7)$$

$$f, \ D_t^{2d+4} f \in L^2(0, 2T; \ W_2^2(\mathcal{D})), \ f\Big|_{t=0} = \nu(x), \ \min_{\bar{\mathcal{D}}} |\nu(x)| > 0,$$

where d is natural.

It may be shown that inverse problems 1, 2 are equivalent to the problem

$$-\lambda_k^2 u_k - A u_k - a u_k = \frac{f_k}{v} \left[-\sum_{k=0}^{\infty} \lambda_k^2 u_k - A\tau - a\tau \right], \quad k = 0, 1, \ldots,$$

$$-\lambda_k^2 v_k - A v_k - a v_k = \frac{g_k}{\nu} \left[-\sum_{k=1}^{\infty} \lambda_k^2 u_k - A\tau - a\tau \right], \quad k = 1, 2, \ldots, \quad (2.8)$$

$$u_k\Big|_S = v_k\Big|_S = 0.$$

Theorem 2.4. Let conditions (2.32) hold.

1. Then the inverse problem 1 has no more than one solution if

$$-\lambda_k^2 + z_n^2 \neq 0 \qquad \forall k = 0, 1, \ldots; \qquad n = 1, 2, \ldots, \tag{2.9}$$

$$\left(A\varphi_n + a\varphi_n = -z_n^2 \varphi_n, \quad \varphi_n \Big|_S = 0, \quad \int_{\mathcal{D}} \varphi_n \varphi_m dx = \delta_{nm} \right).$$

2. Let condition (2.35) hold and

$$f(x,t) = \sum_{k=0}^{\infty} f_k \cos \lambda_k t + \sum_{k=0}^{\infty} g_k(x) \sin \lambda_k t \equiv F(t) + g(t,x), \quad F(0) \neq 0.$$

Then the inverse problem 1 has no more than one solution if

$$\int_0^{2T} F(t) \cos(T-t) z_n dt \neq 0 \qquad \forall n = 1, 2, \ldots. \tag{2.10}$$

3. Let condition (2.35) hold and

$$\nabla_x f_k = 0, \quad \nabla_x g_k = 0, \quad k = 0, 1, 2, \ldots.$$

Then the inverse problem 1 has no more than one solution if and only if (2.36) holds.

4. Let conditions (2.33) hold and

$$\frac{2T^{4d-1} c_1}{\pi^{4d+1} \delta^2 c^2} \int_0^{2T} \left\| \mathcal{D}_t^{2d+4} f \right\|_0^2 dt \left[\frac{3}{2} R^4 + c^2 T^4 \right] < \frac{1}{2}.$$

Then there exists a unique solution to inverse problem 2 and

$$\int_{\mathcal{D}} \sum_{k=1}^{\infty} \lambda_k^8 \left[\left(|\Delta u_k|^2 + |\Delta v_k|^2 \right) + \lambda_k^4 (u_k^2 + v_k^2) \right] d\mathcal{D}$$

$$\leq 4 \left\| \frac{\Delta \tau}{v} \right\|_C^2 \left(2T^4 + \frac{3R^4 T^{4d-5}}{c_1^2 \pi^{4d}} \right) \int_0^{2T} \left\| D_t^{2d+4} f \right\|_0^2 dt.$$

Here c_1 is such a constant that $\left\| u \right\|_0^2 \leq c_1 \left\| \Delta u \right\|_0^2, \; u \Big|_S = 0.$

Proof. Let such integers k_0, n_0 exist that

$$-\lambda_{k_0}^2 + z_{n_0}^2 = 0.$$

Then the functions $u_k = 0$, $k = 0, 1, \ldots$ and $v_{k_0} - Av_{k_0} - av_k = z_{n_0}^2 v_{k_0}$, $v_{k_0}\Big|_S = 0$, $v_k = 0$, $k \neq k_0$ solve homogeneous problem (2.34).

2. Note that, if for certain $k = \{k_1, k_2, \ldots\} = M$, $f_k = 0$, then the first equation (2.34) yields $u_k = 0$, $k = M$. Therefore, we shall consider the first equation from (2.34) for $k \notin M$. As the system $\{\varphi_k\}$ is complete in $W_2^2(\mathcal{D}) \bigcap \overset{0}{W}{}_2^1(\mathcal{D})$, then

$$u_k = \sum_{n=1}^{\infty} u_{kn}\varphi_n(x).$$

Hence and from the first equation (2.34) for $\tau = 0$, it follows that

$$u_{kn} = -\frac{æ_n f_k}{F(0)(z_n^2 - \lambda_k^2)}, \quad k \notin M, \quad æ_n \sum_{k=0}^{\infty} \frac{f_k}{\lambda_k^2 - z_n^2} = 0, \quad æ_n = \sum_{l=1}^{\infty} \lambda_l^2 u_{ln}. \tag{2.11}$$

As

$$f_k = \frac{1}{T} \int_0^{2T} F(t) \cos \lambda_k t dt, \quad k = 1, 2, \ldots,$$

then (2.35) yields

$$z_n æ_n \sum_{k=0}^{\infty} \frac{f_k}{\lambda_k^2 - z_n^2} = -\frac{æ_n}{\sin T z_n} \int_0^{2T} F(t) \cos(T - t) z_n dt.$$

From (2.37), if

$$\int_0^{2T} F(t) \cos(T - t) z_n dt \neq 0, \quad n = 1, 2, \ldots,$$

then $æ_n = 0$, $u_{kn} = 0$; therefore, $u_k = 0$. From the second equation (2.34) we obtain that $v_k = 0$, since $\lambda_k^2 \neq z_n^2$.

3. Sufficiency of (2.36) we have proved above. To prove necessity, note that if n_0 is such a number that

$$\int_0^{2T} F(t)\cos(T-t)z_{no}\,dt = 0,$$

then the functions

$$u_k = -\frac{f_k\varphi_{no}}{F(0)(z_{no}^2-\lambda_k^2)}, \qquad v_k = -\frac{g_k\varphi_{no}}{F(0)(z_{no}^2-\lambda_k^2)},$$

solve homogeneous problem (2.34).

To solve (2.34) we may apply the successive approximations method (Dezin, 1980).

Remark 2. Note, that if $\frac{R^2}{T^2}$ is an algebraic number of the degree d, then condition (2.33) holds automatically.

Remark 3. Problems 1, 2 may be considered with other conditions (2.29), in particular, with nonlocal.

Remark 4. Problems 1, 2 simulate, for example, the following problem.

Suppose we have tubing of length R with open ends (it may be that one end is open and the other is closed, which corresponds to the boundary condition $u\Big|_{k=0} = 0$, $u_x\Big|_{x=k} = 0$).

The source $F(x,t) = f(t)h(x)$ is situated somewhere along the tubing ($f(t)$ is a periodic function).

A function $u(x,t)$ will be a pressure in the tubing, so that

$$u_{tt} - c^2u_{xx} = F(x,t), \quad 0 < x < R, \quad |t| < \infty.$$

We are interested to find the stationary oscillating process. The problem is to find such $F(x,t)$ (i.e., h) that in the given moment T we can obtain the required pressure $u^*(x)$.

Inverse problems for hyperbolic equations in the classes of functions decreasing by t, with data inside the domain. In the domain $Q = (-\infty,\infty)\times\mathcal{D}$, where $\mathcal{D}$ is a domain in R^n with a boundary $S \in C^2$, consider the problem

$$Lu = K(x)u_{tt} + \alpha(x)u_t - \sum_{i,j=1}^{n} \left(a_{ij}(x)u_{x_i}\right)_{x_j}$$

$$+ \sum_{i=1}^{n} a_i(x,t)u_{x_i} + a(x)u = f(t,x)h(x), \tag{2.12}$$

$$u\Big|_{\Gamma} = \varphi, \qquad \Gamma = (-\infty, \infty) \times S,$$

$$k \geq \delta > 0, \qquad \sum_{i,j=1}^{\infty} a_{i,j}\xi_i\xi_j \geq \mu_0|\xi|^2, \qquad a_{ij} = a_{ji}. \tag{2.13}$$

Further, we shall assume that the coefficients of (2.38) are sufficiently smooth.

Inverse problem. To find functions u, h (with f given) satisfying (2.38), (2.39) and

$$u\Big|_{t=0} = u_0(x). \tag{2.14}$$

We shall assume that $|f(x,0)| \geq \delta > 0$. It turns out that inverse problem (2.38)-(2.40) in certain classes is equivalent to the Dirichlet problem for the integro-differential equation

$$-Av + av - (\alpha i\lambda + \lambda^2 k)v + \sum_{i=1}^{n} \int_{-\infty}^{+\infty} \hat{a}_i(x, \lambda - s)v_{x_i}(x,s)ds$$

$$= \frac{\hat{f}(x,\lambda)}{f(x,0)} \left[\Phi - \int_{-\infty}^{+\infty} (\alpha i\lambda + k\lambda^2)v d\lambda\right], \tag{2.15}$$

$$v\Big|_{S} = \hat{\varphi}.$$

Solvability of (2.41) may be proved for certain conditions imposed on input data. The essence of these conditions is the proximity of initial data $u_0(x)$.

The functions $u(x,t)$ and $h(x)$, thus, may be found by the formulas

$$h(x) = \frac{1}{f(x,0)}\left[\Phi - \int_{-\infty}^{+\infty}(\alpha i\lambda + k\lambda^2)vd\lambda\right],$$

$$u = \int_{-\infty}^{+\infty} v(x,\lambda)e^{-i\lambda t}d\lambda, \tag{2.16}$$

$$\Phi = -Au_0 + \sum_{i=1}^{n} a_i(x,0)u_{0x} + au_0, \quad Au \equiv \sum_{i,j=1}^{n}\left(a_{ij}(x)u_{x_i}\right)_{x_j},$$

$$\hat{f}(x,\lambda) = \frac{1}{2\pi}\int_{-\infty}^{+\infty} f(x,t)e^{i\lambda t}dt.$$

Now let $f(x,t) \equiv f(t)$, $0 \notin \operatorname{supp} f(t)$, $|\alpha| = \text{const} > 0$, $a_i = \varphi = 0$, $k = 1$.

In this case we can not obtain equation (2.41), since $f(0) = 0$.

The inverse problem solution we seek in the class

$$h \in W_2^2(\mathcal{D}) \bigcap \overset{0}{W}{}_2^1(\mathcal{D}); \quad u, D_t^3 u \in L^2\left(-\infty,\infty; W_2^2(\mathcal{D}) \bigcap \overset{0}{W}{}_2^1(\mathcal{D})\right),$$

$$D_t^4 u \in L^2(Q).$$

Introduce the following notations

$$A\varphi_n - a\varphi_n = -\sigma_n^2\varphi_n, \quad \varphi_n\Big|_S = 0, \quad \int_{\mathcal{D}} \varphi_n\varphi_m dx = \delta_{nm},$$

$$a_{ij} \in C^\infty(\bar{\mathcal{D}}), \quad a \in C^\infty(\bar{\mathcal{D}}), \quad a \geq 0, \quad S \in C^\infty.$$

For definiteness, consider the case $\alpha > 0$, $4\sigma_n^2 - \alpha^2 > 0 \ \forall n$. The other cases $|\alpha| \neq 0$ and $4\sigma_n^2 - \alpha^2 \leq 0$, $n \leq n_0$, and $4\sigma_n^2 - \alpha^2 > 0$, $n > n_0$ are considered analogously.

Denote

$$\omega_n = \left(4\sigma_n^2 - \alpha^2\right)^{1/2}, \quad f_n = \int_{-\infty}^{0} D_t^2\left(f\exp\left(\frac{\alpha t}{2}\right)\right)\sin\frac{t\omega_n}{2}dt,$$

$$K_n = -\frac{\omega_n^3}{8\sigma_n^2}\int_{\mathcal{D}}(Au_0 - au_0)\varphi_n d\mathcal{D}.$$

Theorem 2.5. Let $f(t) \equiv 0$, $t \geq 0$ $t \leq -T_1 < 0$, and for $f_n \neq 0$ the estimates

$$\frac{c_1}{\sigma_n^\beta} \leq |f_n| \leq \frac{c_0}{\sigma_n}, \quad \text{where } \beta \text{ is natural,}$$

$$u_0 \in W_2^{\beta+5}(\mathcal{D}) \bigcap \overset{0}{W}{}_2^{\beta+4}(\mathcal{D}); \quad f, D_t^2 f \in L^2(-\infty, 0)$$

hold. Then

1. There exists a unique solution to (2.38)-(2.40) if and only if

$$f_n \neq 0, \qquad n \in \{n_1, n_2, \ldots\} = M,$$

and

$$\int_{-\infty}^{+\infty} \left(\left\|u\right\|^2_{W_2^2} + \left\|D_t^3 u\right\|^2_{W_2^2} + \left\|D_t^4 u\right\|^2_0 \right) dt$$

$$\leq \frac{c_2}{c_1} \left(\left\|Au_0 - au_0\right\|^2_{W_2^{\beta+3}} + \left\|f\right\|^2_{W_2^2} \right),$$

$$\left\|h\right\|^2_{W_2^2} \leq \frac{c_2}{c_1} \left(\left\|Au_0 - au_0\right\|^2_{W_2^\beta} \right).$$

2. If $f_n = K_n = 0$, $n \in M$, $f_n \neq 0$, $K_n \neq 0$, $n \notin M$, then all solutions to the inverse problem have the form $h(x) = \psi(x) + F(x)$, $u = u_1 + u_2$, F, u_2 are uniquely determined by f, u_0 and the estimates from p.1 hold. Here $\psi \in H_M$, $u_1 \in V_M$ are arbitrary functions.

3. If $f_n = K_n = 0$ $\forall n = 1, 2, \ldots$, then the inverse problem has an infinite number of solutions $h = \psi \in H_\infty$, $u = u_1 + u_2$, $u_1 \in H_\infty$. For u_2 the estimate from p.1 holds.

4. If $f_n \neq 0$, $K_n = 0$ $\forall n = 1, 2, \ldots$, then inverse problem (2.38)-(2.40) has a unique solution $h = 0$.

5. If for a certain $n_1 : f_{n_1} = 0$, $K_{n_1} \neq 0$, then there are no solutions to inverse problem (2.38)-(2.40).

Here, we have denoted

$$H_M = \Big\{\psi : \psi \in W_2^2(\mathcal{D}) \bigcap \overset{0}{W}{}_2^1(\mathcal{D});\ \ \psi = \sum_{n \in M} \overset{0}{\gamma}_n \varphi_n,\ \ \overset{0}{\gamma}_n \in R\Big\},$$

$$V_M = \Big\{u_1(x,t) : \int\limits_{-\infty}^{+\infty} u_1(x,t) e^{i\lambda t} dt = v_1(x,\lambda),\ \ v_1\Big|_S = 0,$$

$$v_1 = \sum_{n \in M} \frac{\hat{f}(\lambda) \overset{0}{\gamma}_n \psi_n}{\sigma_n^2 - \lambda^2 - i\lambda\alpha},\ \ \int\limits_{-\infty}^{+\infty} (1 + |\lambda|^6) \Big\| v_1 \Big\|_{W_2^2}^2 d\lambda < \infty, \Big\},$$

$$H_\infty = \Big\{\psi : \psi \in W_2^2(\mathcal{D}) \bigcap \overset{0}{W}{}_2^1(\mathcal{D}),\ \ \psi = \sum_{n \in M} \overset{0}{\gamma}_n \varphi_n\Big\}.$$

So, in this case, we found the necessary and sufficient solvability conditions for (2.38)-(2.40) and have described the solution kernel of the inverse problem. This kernel, depending on the domain geometry and input data may be both finite-dimensional and infinite-dimensional.

Proof. We shall give the scheme of the proof. Inverse problem (2.38)-(2.40) is equivalent to the equation system

$$-\lambda^2 v - Av - i\lambda\alpha v + av = h(x)\hat{f}(\lambda),\ \ x \in \mathcal{D},\ \ v\Big|_S = 0,$$

$$-\frac{1}{2\pi} \int\limits_{-\infty}^{+\infty} (\lambda^2 + i\lambda\alpha) v(\lambda, x) d\lambda = Au_0 - au_0,\quad x \in \mathcal{D}. \tag{2.17}$$

We shall look for the solution for this system in the form

$$v(x,\lambda) = \sum_{n=1}^{\infty} \varphi_n(x) c_n(\lambda), \qquad h(x) = \sum_{n=1}^{\infty} \gamma_n(x) \varphi_n(x)$$

since the system $\{\varphi_n\}$ is complete in $W_2^l(\mathcal{D}) \cap \overset{0}{W}{}_2^1(\mathcal{D})$.

Hence and from (2.43), applying the residual theory, we obtain

$$-\frac{8\sigma_n^2}{\omega_n^3} \gamma_n \int\limits_{-\infty}^{0} D_t^2 \left(f \exp\left(\frac{\alpha t}{2}\right)\right) \sin \frac{t\omega_n}{2} dt = \int\limits_{\mathcal{D}} \Big(Au_0 - au_0\Big) \varphi_n dx,$$

$$c_n(\lambda) = \frac{\hat{f}(\lambda)\gamma_n}{\sigma_n^2 - \lambda^2 - i\lambda\alpha}. \tag{2.18}$$

The analysis of (2.44) leads to the result.

It should be noted that the equalities

$$\int_0^T D_t^2 \left(f \exp\left(\frac{\alpha t}{2}\right)\right) \sin \frac{t\omega_n}{2} dt = 0,$$

which hold for $n \in N$ do not imply that

$$D_t^2 \left(f \exp\left(\frac{\alpha t}{2}\right)\right) = 0.$$

Really, let $Au = u^N$, $4a = \alpha^2 = \text{const}$, $\mathcal{D} = (0, l)$. Then

$$\sigma_k^2 = \frac{\pi^2 k^2}{l^2} + a, \quad \varphi_k = \sin \frac{\pi k x}{l}, \quad \omega_k = \frac{2\pi k}{l}, \quad \frac{\omega_k}{\pi k} = \frac{2}{l},$$

$$D_t^2 \left(f \exp\left(\frac{\alpha t}{2}\right)\right) = t(T-t) \sin \frac{\pi t}{T},$$

T/l is even, and, thus,

$$\int_0^{T/2} t(T-2t) \sin \frac{2\pi k}{T} \sin \left(t\pi k \frac{\omega_k}{\pi k}\right) dt \equiv 0,$$

for $k \in N$.

Inverse problems for hyperbolic equations with data in the boundary of the domain. The inverse problem with the additional information given inside the domain sometimes is not satisfactory from the practical point of view. In this section we shall remove these conditions.

A. Let $\mathcal{D} = \{(\bar{x}, z);\ x \in R^{n-1},\ z = 0\}$. In the domain $Q = (-\infty, \infty) \times \mathcal{D}$ consider the problem

$$Lu = u_{tt} + \alpha u_t - u_{zz} - \Delta_x u + au = f(t, x) h(x, z), \tag{2.19}$$

$$u\Big|_{z=0} = \varphi_0(x,t),$$

$$\alpha = \text{const} \neq 0, \qquad a = \text{const} > 0. \tag{2.20}$$

Inverse problem. To find u, h (with f given) satisfying (2.45), (2.46) and

$$u_z(x,0,0) = \varphi_1(x). \tag{2.21}$$

The solution to (2.45)-(2.47) will be taken from the class H: smooth u, decreasing as $|x|^2 + z^2 + t^2 \to \infty$ and

$$\frac{\partial}{\partial z}\left[\frac{1}{\gamma(z)}\left(\Delta_x u\Big|_{t=0} + u_{zz}\Big|_{t=0} - au_0\Big|_{t=0}\right)\right] = 0, \tag{2.22}$$

with γ given, $|\gamma(z)| > 0$, $|z| < \infty$, $|\gamma(z)| \to 0$ as $z \to \infty$.

Then in this class, inverse problem (2.43)-(2.47), if $|\varphi(0,z)| \geq \delta > 0$, is equivalent to the system

$$-v_{zz} - \Delta_x v + av - (\lambda + i\lambda\alpha)v = \frac{\hat{f}(\lambda,z)}{f(0,z)}\Bigg[-\int\limits_{-\infty}^{+\infty}(\lambda^2 + i\lambda\alpha)v\,d\lambda$$

$$-\Delta_x u_0 - u_{0zz} - au_0\Bigg], \quad x \in R^{n-1}, \quad z > 0, \quad v\Big|_{z=0} = \hat{\varphi}_0(\lambda, x), \tag{2.23}$$

$$\Delta_x u_0 + u_{0zz} - au_0 = \gamma(z)b(x), \quad x \in R^{n-1}, \quad z > 0, \tag{2.24}$$

$$u_0 = \int\limits_{-\infty}^{+\infty} v\,d\lambda, \quad u_0\Big|_{z=0} = \varphi_0(x,0), \quad u_{0z}\Big|_{z=0} = \varphi_1(x), \tag{2.25}$$

where $v, u_0, b(x)$ are the desired functions.

The results obtained before give the necessary and sufficient solvability conditions for the Cauchy problem for elliptic equation (2.50), from which follows that

$$b(x) = \int_{-\infty}^{+\infty} \left[\hat{\varphi}_0(\xi, 0) + \frac{\hat{\varphi}_1(\xi)}{\sqrt{a + |\xi|^2}} \right] \exp\Big(-i(\xi, x)\Big) d\xi$$

$$\times \left(\int_0^{\infty} \gamma(z) \exp\left(-z\sqrt{a + |\xi|^2}\right) dz \right)^{-1}. \tag{2.26}$$

Thus, problem (2.49)-(2.51), taking account of (2.52), is reduced to the Dirichlet problem for the integro-differential equation ($x \in R^{n-1}$, $z > 0$)

$$-v_{zz} - \Delta_x v + av - (\lambda^2 + i\lambda\alpha)v = -\frac{\hat{f}(\lambda, z)}{f(0, z)} \left\{ \int_{-\infty}^{+\infty} (\lambda^2 + i\lambda\alpha) v d\lambda \right.$$

$$-\gamma(z) \int_{-\infty}^{+\infty} \left[\hat{\varphi}_0(\xi, 0) + \frac{\varphi_1(\xi)}{\sqrt{a + |\xi|^2}} \right] \exp\Big(-i(\xi, x)\Big)$$

$$\left. \times \left(\int_0^{\infty} \gamma(z) \exp\left(-z\sqrt{a + |\xi|^2}\right) dz \right)^{-1} d\xi \right\}.$$

$$v\Big|_{z=0} = \hat{\varphi}_0(\lambda, x). \tag{2.27}$$

After solving (2.53), the functions u, h are obtained by the formulas

$$u = \int_{-\infty}^{+\infty} v(\lambda, x, z) e^{-i\lambda t} d\lambda,$$

$$h(x, z) = \frac{1}{f(0, z)} \left\{ -\int_{-\infty}^{+\infty} (\lambda^2 + i\lambda\alpha) v d\lambda - \gamma(z) \int_{-\infty}^{+\infty} \left[\hat{\varphi}_0(\xi, 0) + \frac{\varphi_1(\xi)}{\sqrt{a + |\xi|^2}} \right] \right.$$

$$\left. \times \exp\Big(-i(\xi, x)\Big) \left(\int_0^{\infty} \gamma(z) \exp\left(-z\sqrt{a + |\xi|^2}\right) dz \right)^{-1} d\xi \right\}.$$

where

$$\hat{\varphi}_0(\xi, 0) = \frac{1}{2\pi} \int_{-\infty}^{+\infty} \varphi_0(x, 0) e^{i(\xi, x)} d\xi \quad \hat{\varphi}_1(\xi) = \frac{1}{2\pi} \int_{-\infty}^{+\infty} \varphi_1(x) e^{i(\xi, x)} dx,$$

$$\hat{f}(\lambda, z) = \frac{1}{2\pi} \int_{-\infty}^{+\infty} f(t, z) e^{i\lambda t} dt, \quad |\xi|^2 = \xi_1^2 + \xi_2^2 + \ldots + \xi_n^2.$$

B. In the domain $\{t > 0,\ x \in R^n,\ z > 0\}$, consider the problem

$$Lu \equiv u_{tt} - u_{zz} - \Delta_{x'} u - u_{x_n x_n} = h(z, x', t) f(z, x', z_n, t), \tag{2.28}$$

$$u\Big|_{t=0} = \varphi_0, \quad u_t\Big|_{t=0} = \varphi_1, \quad u_z\Big|_{z=0} = 0, \tag{2.29}$$

$$x = (x', x_n), \qquad x' \in R^{n-1}.$$

Inverse problem. To find functions u, h (with f given) satisfying (2.54), (2.55), and

$$u(0, x', 0, t) = \varphi_2(x', t) \tag{2.30}$$

in the class H_1: the smooth functions u, decreasing as $|x|^2 + z^2 \to \infty$, and

$$\frac{\partial}{\partial z} \left[\frac{1}{\gamma(z)} \Big(u_{tt}(z, x', 0, t) - u_{zz}(z, x', 0, t) - \Delta_{x'}(z, x', 0, t) \Big) \right] = 0,$$

with γ given, or, in the class H_2: the smooth functions u, decreasing as $|x|^2 + z^2 \to \infty$, and

$$\frac{\partial}{\partial z} \left[\frac{1}{\gamma(t)} \Big(u_{tt}(z, x', 0, t) - u_{zz}(z, x', 0, t) - \Delta_{x'}(z, x', 0, t) \Big) \right] = 0,$$

and with γ again given.

Inverse problem (2.53)-(2.56) in the class H_1 is equivalent to solvability of the problem

$$v_{tt} - v_{zz} - \Delta_{x'} v + \lambda^2 v = \frac{\hat{f}(z, x', \lambda, t)}{f(z, x', 0, t)} \left[\int_{-\infty}^{+\infty} \lambda^2 v d\lambda + u_{0tt} - u_{0zz} - \Delta_{x'} u_0 \right], \tag{2.31}$$

$$v\Big|_{t=0} = \hat{\varphi}_0, \quad v_t\Big|_{t=0} = \hat{\varphi}_1, \quad \left(t > 0,\ z > 0,\ x' \in R^{n-1}\right),$$

$$u_{0tt} - u_{0zz} - \Delta_{x'} u_0 = b(t, x')\gamma(z), \quad t > 0,\ z > 0,\ x' \in R^{n-1},$$

$$u_0\Big|_{z=0} = \varphi_2(x', t), \quad u_{0z}\Big|_{z=0} = 0, \quad u_0\Big|_{t=0} = \varphi_0(z, x', 0, 0),$$

$$u_{0t}\Big|_{t=0} = \varphi_1(z, x', 0, 0), \tag{2.32}$$

and in the class H_2, of the problem

$$w_{tt} - w_{zz} - \Delta_{x'} w + \lambda^2 w = \frac{\hat{f}(z, x', \lambda, t)}{f(z, x', 0, t)} \left[\int_{-\infty}^{+\infty} \lambda^2 w d\lambda + u_{0tt} - u_{0zz} - \Delta_{x'} u_0 \right], \tag{2.33}$$

$$w\Big|_{t=0} = \hat{\varphi}_0, \quad w_t\Big|_{t=0} = \hat{\varphi}_1,$$

$$u_{0tt} - u_{0zz} - \Delta_{x'} u_0 = a(x', z)\gamma(t),$$

$$u_0\Big|_{z=0} = \varphi_2(x', t), \quad u_{0z}\Big|_{z=0} = 0, \quad u_0\Big|_{t=0} = \varphi_0(z, x', 0, 0),$$

$$u_{0t}\Big|_{t=0} = \varphi_1(z, x', 0, 0), \tag{2.34}$$

In the system (2.57), (2.58) the unknown functions are v, u, b. In the system (2.59), (2.60) the unknown functions are w, u_0, a, where

$$\hat{f}(z, x', \lambda, t) = \frac{1}{2\pi} \int_{-\infty}^{+\infty} \hat{f}(z, x', x_n, t) e^{i\lambda x_n} dx_n$$

and $\hat{\varphi}_0, \hat{\varphi}_1$ are the Fourier transforms of φ_0, φ_1, by the variable x_n.

Problem (2.60) was investigated in the previous section, where the necessary and sufficient conditions for its solvability were obtained. They follow from solvability of the Cauchy problem for an elliptic equation in the classes of finite smoothness functions. So, problem (2.59) is the Cauchy problem for the integro-differential equation, the solvability of which was investigated by Bubnov (1987b).

Solvability of problems (2.58), (2.59) will be considered below.

Problems with moving sources and their numerical solution. In the domain $Q = (0,T) \times \mathcal{D} \times (0,\pi)$, $z \in (0,\pi)$, $x \in \mathcal{D}$, $Q_0 = \mathcal{D} \times (0,\pi)$, where $\mathcal{D}$ is a domain in R^n with a boundary $S \in C^4$, consider the problem

$$Lu \equiv u_{tt} - u_{zz} - A(x,\mathcal{D})u + a(x,t)u = h(x,t)f(t,x,z), \tag{2.35}$$

$$u\Big|_{t=0} = u_0, \quad u_t\Big|_{t=0} = u_1, \quad u_z\Big|_{z=0} = u_z\Big|_{z=\pi} = 0, \quad \frac{\partial u}{\partial N}\Big|_{\Gamma_1} = 0, \tag{2.36}$$

$$Au \equiv \sum_{i,j=1}^{n} \left(a_{ij}(x)u_{x_i}\right)_{x_j}, \quad a_{ij} = a_{ji} \in C^4(\bar{\mathcal{D}}), \quad \sum_{i,j=1}^{n} a_{ij}\xi_i\xi_j \geq \mu_0|\xi|^2,$$

$\frac{\partial u}{\partial N} = \sum_{i,j=1}^{n} a_{ij}u_{x_i}n_j$, $n = (n_1,\ldots,n_n)$ is the exterior normal to S, $\Gamma_1 = (0,T) \times S$.

Inverse problem. To find functions u, h (with f given) satisfying (2.61), (2.62) and

$$u\Big|_{z=0} = 0. \tag{2.37}$$

A. Let $|f(t,x,0)| \geq \delta > 0$; then inverse problem (2.61)-(2.63) in the class of functions ($Q_1 = (0,T) \times \mathcal{D}$)

$$h \in W_2^2(Q_1), \quad u \in W_2^2(Q), \quad D_t^3 u_{tt}, \quad D_z^4 u_t, \quad D_z^4 \nabla_x u, \quad D_z^3 Au \in L^2(Q) \tag{2.38}$$

is equivalent to the first initial boundary-value problem for the infinite system of hyperbolic equations

$$u_{ktt} - Au_k + \lambda_k^2 u_k + a(x,t)u_k = \frac{f_k(x,t)}{f(t,x,0)} \sum_{n=1}^{\infty} \lambda_n^2 u_n, \tag{2.39}$$

$$u_k\Big|_{\Gamma} = 0, \quad u_k\Big|_{t=0} = u_{0k}, \quad u_{kt}\Big|_{t=0} = u_{1k},$$

where

$$f_k = \frac{2}{\pi}\int_0^{\pi} f(t,x,z)\cos kz dz, \quad u_{ik} = \frac{2}{\pi}\int_0^{\pi} u_i \cos kz dz, \quad \lambda_k = k.$$

Theorem 2.6. Let the concordance conditions for u_0, u_1 hold and

$$u_0, Au_0, D_z^3 Au_0 \in L^2\big(0,\pi; W_2^2(\mathcal{D})\big), \quad D_z^5 u_0 \in L^2(Q_0);$$

$$D_z^k u_0\Big|_{z=0} = D_z^k u_0\Big|_{z=\pi} = 0, \ \ k = \overline{1,3}, \nabla_x u_1, u_1, D_z^5 u_1, D_z^3 Au_2 \in L^2(Q_0);$$

$$D_z^k u_1\Big|_{z=0} = D_z^k u_1\Big|_{z=\pi} = 0, \ \ k = \overline{1,3} \ \ f\Big|_{z=0} = f_z\Big|_{z=\pi} = 0,$$

$$f, D_z^4 f \in L^2(Q), \quad D_z^3 f, D_z^3 f_{tt} \in L^2\big(0,\pi; W_2^2(Q_1)\big);$$

$$D_t^3 f(t,x,0), f(t,x,0) \in L^2\big(0,T; W_2^2(\mathcal{D}) \bigcap C(\bar{\mathcal{D}})\big).$$

Then there exists a unique solution to (2.65) and

$$\int_{\mathcal{D}} \sum_{k=0}^{\infty} \left(1+k^6\right) \left[u_{ktt}^2 + \left|\nabla_x u_{kt}\right|^2 + \left(1+k^2\right)\left(\left|\nabla u_k\right|^2 + u_{kt}^2\right)\right] d\mathcal{D}$$

$$+ \int_{\mathcal{D}} \sum_{k=0}^{\infty} \left(1+k^6\right) u_{kx_ix_j}^2 \, d\mathcal{D} \le c(f, u_0, u_1),$$

$$c(\xi,\eta,\lambda) \to 0, \quad \xi^2 + \eta^2 + \lambda^2 \to 0.$$

Proof of this theorem may be done by the successive approximations method (Dezin, 1980).

Corollary. Let conditions of theorem 2.6 hold. Then problem (2.61)-(2.63) is uniquely solvable and

$$u = \sum_{k=0}^{\infty} u_k \cos\, kz, \qquad h(x,t) = \frac{1}{f(t,x,0)} \sum_{k=0}^{\infty} k^2 u_k(x,t);$$

$$f(t,x,z) = f(z), \quad z \in (0,2\pi), \ \ x \in \mathcal{D}, \ \ t \in (0,\infty).$$

Inverse problem. To find functions u, h such that

$$u_{tt} - u_{zz} - A(x, \mathcal{D})u + a(x)u = h(x,t)f(z), \tag{2.40}$$

$$u\Big|_{t=0} = u_t\Big|_{t=0} = 0, \quad D_z^k u\Big|_{z=0} = D_z^k u\Big|_{z=2\pi} = 0, \quad k = 0, 1,$$

$$\frac{\partial u}{\partial N}\Big|_{\Gamma_1} = \sum_{i,j=1}^{n} a_{ij} u_{x_i} n_j\Big|_{\Gamma_1} = 0, \quad \Gamma_1 = (0, \infty) \times \partial\mathcal{D} \tag{2.41}$$

(n is the external normal vector to $\partial\mathcal{D}$, the operator A is determined as in (2.62)) and, in addition,

$$u\Big|_{z=0} = \varphi_0(x,t). \tag{2.42}$$

Theorem 2.7. Let the following conditions hold:

$$f(z) \in \overset{0}{W}{}_2^{\beta}(0, 2\pi), \quad \beta > 2, \quad a_{ij}, a \in C^{|\alpha|}(\bar{\mathcal{D}}); \quad a > 0,$$

$$\frac{1}{2\pi}\left|\int_0^{2\pi} f(z)dz\right| - \frac{4}{\sqrt{p_0}(1 - e^{-\pi p_0})}\left(\int_0^{2\pi} |f'|^2 dt\right)^{1/2} \geq \delta > 0,$$

let φ_0 be such that

$$\frac{\partial \varphi_0}{\partial N}\Big|_S = 0, \quad \varphi_0\Big|_{t=0} = \varphi_{0t}\Big|_{t=0} = 0, \quad \varphi_{0tt} - A\varphi_0 + a\varphi_0 \equiv \Phi,$$

$$\Phi\Big|_{t=0} = D_t\Phi\Big|_{t=0} = \ldots = D_t^{\gamma}\Phi\Big|_{t=0} = 0, \quad \frac{\partial \Phi}{\partial N}\Big|_S = 0,$$

let γ be natural and greater than 1; and

$$\int_0^{\infty} e^{-2p_0 t}\left[\left\|\Phi\right\|^2_{W_2^{|\alpha|}(\mathcal{D})} + \left\|\Phi_t\right\|^2_{W_2^{|\alpha|}(\mathcal{D})} + \left\|D_t^{\gamma+1}\Phi\right\|^2_{L^2}\right] dt < \infty.$$

Then there exists a unique solution to inverse problem (2.66)-(2.68) from the class

$$h, D_t^\gamma h, D_x^\alpha h \in L^\infty\left(0,T;L^2(\mathcal{D})\right) \quad \forall T>0, \quad |\alpha|>2, \quad \frac{\partial h}{\partial N}\Big|_S = 0;$$

$$h\Big|_{t=0} = \ldots = D_t^\gamma h\Big|_{t=0} = 0, \quad \int_0^\infty e^{-2p_0 t} h^2 dx dt < \infty$$

$$u, \; D_x^\beta D_x^\alpha u, \; D_x^{\alpha_1} D_z^{\beta_1} u_{tt} \in L^\infty\left(0,T;L^2(\mathcal{D})\right) \quad \forall T>0;$$

$$|\alpha_1| \le |\alpha| - 2, \qquad \beta_1 < \beta - 2;$$

$$\int_0^\infty e^{-2p_0 t}\left[\left|u\right|^2 + \left|u_{tt}\right|^2 + \left|D_x^\beta D_x^\alpha u\right|^2\right] dx dz dt < \infty, \quad p_0 > 0.$$

Proof. We shall outline the scheme of the proof. As the system $\{1, \cos z, \ldots, \cos nz, \ldots, \sin z, \ldots, \sin nz, \ldots\}$ is complete in $L^2(0,2\pi)$ then

$$f(z) = \sum_{n=0}^\infty \left(f_n \cos nz + g_n \sin nz\right).$$

We shall look for the solution to (2.66)-(2.68) in the form

$$u = \sum_{n=0}^\infty \left[u_n(x,t)\cos nz + v_n(x,t)\sin nz\right], \quad h(x,t) = \sum_{n=0}^\infty b_n(t)\varphi_n(x),$$

$$A\varphi_n + a\varphi_n = -\lambda_n^2 \varphi_n; \quad \frac{\partial \varphi_n}{\partial N}\Big|_S = 0, \quad \int_{\mathcal{D}} \varphi_n \varphi_m dx = \delta_{nm}.$$

Hence, to determine $b_n(t)$, we obtain the relation

$$-\sum_{k=1}^\infty \frac{f_k k^2}{\sqrt{k^2+\lambda_n^2}} \int_0^1 b_n(\tau)\sin(t-\tau)\sqrt{k^2+\lambda_n^2}\, d\tau = \Phi_n(t), \tag{2.43}$$

$$\Phi_n = \int_{\mathcal{D}} \Phi \varphi_n dx.$$

From (2.69), doing the Laplace transformation by t, for Re$p \geq p_0$, we obtain

$$\frac{\hat{b}_n(p)}{2\pi}\int_0^{2\pi}\Bigg[f(z) - \frac{f'(z)}{\exp\big(\pi\sqrt{p^2+\lambda_n^2}\big) - \exp\big(-\pi\sqrt{p^2+\lambda_n^2}\big)}$$

$$\times\Big(\exp\big(-\pi\sqrt{p^2+\lambda_n^2}(\pi - z)\big) - \exp\big(\pi\sqrt{p^2+\lambda_n^2}(\pi - z)\big)\Big)\Bigg]dz$$

$$\equiv \frac{\hat{b}_n(p)}{2\pi}I = \hat{\Phi}_n(p). \tag{2.44}$$

In (2.70), the branch is chosen so that $\mathrm{Re}(\sqrt{p^2+\lambda_n^2}) > 0$.

On the basis of theorem 2.7 conditions imposed on $f(z)$, it follows that the integral I from the left side of (2.70) may be estimated from below $|I| \geq \delta$.

Now, doing the inverse Laplace transformation by p, we obtain the estimates for $b_n(t)$ and, thus, for $h(x,t)$.

The function u may be restored by the formulas

$$u = \tilde{u} + \tilde{v}, \quad \tilde{u} = \sum_{n=0}^{\infty} u_n(x,t)\cos nz, \quad u_k(x,t) = \sum_{n=0}^{\infty} c_n^k(t)\varphi_k(x),$$

$$c_n^k(t) = \frac{f_k}{\sqrt{k^2+\lambda_n^2}}\int_0^t b_n(\tau)\sin(t-\tau)\sqrt{k^2+\lambda_n^2}\,d\tau.$$

The analogous formulas hold for function $\tilde{v}$.

B. In the domain $Q = (a,b)\times R\times(0,T)$ consider the problem

$$u_{tt} - u_{xx} - u_{zz} = h(x,t)f(t,x,z), \quad t\in(0,T), \quad x\in(a,b), \quad z\in R, \tag{2.45}$$

$$u\Big|_{t=0} = 0, \quad u_t\Big|_{t=0} = 0, \quad u\Big|_{x=a} = u\Big|_{x=b} = 0. \tag{2.46}$$

Inverse problem. To find functions u, h (with f given) satisfying (2.71), (2.72), and

$$u\Big|_{z=0} = \varphi_0(x,t). \tag{2.47}$$

If $|f(t,x,0) \geq \delta > 0$, then inverse problem (2.71)-(2.73) is equivalent to the problem

$$v_{tt} - v_{xx} + \lambda^2 v = F_0 + F_1 \int\limits_{-\infty}^{+\infty} \lambda^2 v d\lambda, \quad (t,x) \in (0,T) \times (a,b) = Q_0,$$

$$v\Big|_{t=0} = v_t\Big|_{t=0} = 0, \quad v\Big|_{x=a} = v\Big|_{x=b} = 0, \tag{2.48}$$

$$F_0 = \frac{\hat{f}(t,x,\lambda)}{f(t,x,0)}\Big(\varphi_{0tt} - \varphi_{0xx}\Big), \qquad F_1 = \frac{\hat{f}(t,x,\lambda)}{f(t,x,0)}$$

and the solution u is as follows

$$u = \int\limits_{-\infty}^{+\infty} v(t,x,\lambda) e^{-i\lambda z} d\lambda, \quad h = \frac{1}{f(t,x,0)} \left[\varphi_{0tt} - \varphi_{0xx} + \int\limits_{-\infty}^{+\infty} \lambda^2 v d\lambda\right]. \tag{2.49}$$

Theorem 2.8. Let $\varphi_0 \in W_2^4(Q_0)$, $f_0 = f(t,x,0) \in W_2^2(Q_0)$; $f, D_y^{p/2} f, D_t^{p/2} f \in L^2(Q)$.

Then there exists a unique solution to (2.74) from the class of functions v such that

$$\int\limits_{-\infty}^{+\infty} \int\limits_{a}^{b} (1+\lambda^2)^p \Big[\Big\|v_{tt}\Big\|^2 + \Big\|v_{xt}\Big\|^2 + (1+\lambda^2)|v_t|^2 + |v|^2$$

$$+|v_{xx}|^2\Big] d\lambda dx \leq c(f,\varphi_0), \quad p > \frac{5}{2}.$$

Proof of this theorem may be done by the Halyerkin method. Parallel with problem (2.74), consider the following problem

$$v_{tt}^R - v_{xx}^R + \lambda^2 v^R = F_0 + F_1 \int_{-R}^{R} \lambda^2 v^R d\lambda, \quad (x,t) \in Q_0, \tag{2.50}$$

$$v^R\Big|_{t=0} = 0, \quad v_t^R\Big|_{t=0} = 0, \quad v^R\Big|_{x=a} = v^R\Big|_{x=b} = 0.$$

Theorem 2.8 holds for problem (2.76). Denote $w = v - v^R$; then

$$w_{tt} - w_{xx} + \lambda^2 w = F_1 \int_{-R}^{R} w\lambda^2 d\lambda + F_1 \int_{-\infty}^{-R} \lambda^2 v d\lambda + F_1 \int_{R}^{\infty} \lambda^2 v d\lambda, \tag{2.51}$$

$$w\Big|_{t=0} = w_t\Big|_{t=0} = 0, \quad w\Big|_{x=a} = w\Big|_{x=b} = 0.$$

The estimate from theorem (2.8) and (2.77) yields

$$\frac{d}{dt} \int_{-R}^{R} \int_{a}^{b} (1+\lambda^2)^p \left[|w_t|^2 + |w_x|^2 + \lambda^2 |w|^2\right] d\lambda dx$$

$$\leq c \int_{-R}^{R} \int_{a}^{b} (1+\lambda^2)^p \left[|w_t|^2 + |w_x|^2 + \lambda^2 |w|^2\right] d\lambda dx + \frac{c_1}{(1+R)^{2p-5}}.$$

Hence, follows that

$$\int_{-R}^{R} \int_{a}^{b} (1+\lambda^2)^p \left[|v_t - v_t^R|^2 + |v_x - v_x^R|^2 + \lambda^2 |v - v^R|^2\right] d\lambda dx$$

$$\leq \frac{c_2(e^{Rt} - 1)}{(1+R)^{2p-5}}, \tag{2.52}$$

where c, c_1, c_2 are constants that depend on p, F_0, F_1.

Lemma 2.3. Let conditions of theorem 2.8 hold. Then solutions to problems (2.74), (2.76) satisfy (2.78).

Approximate problem (2.77) in the domain

$$\Omega_{\lambda_R,\tau,h} = \left\{ v_{ij}^n = v(x_i, \lambda_n, t_h) : \left| \begin{array}{ll} x_i = a + (i-1)h, & i = \overline{1, k+1}, \\ \lambda_j = -R + (j-1)R, & j = \overline{1, N+1}, \\ t_n = (n-1)\tau, & n = \overline{1, M+1}, \end{array} \right. \right\}$$

where $h = \frac{b-a}{k}$, $\lambda_R = \frac{2R}{N}$, $\tau = \frac{I}{M}$ by means of the finite difference scheme

$$\left\{ \begin{array}{l} \frac{v^{n+1} - 2v^n + v^{n-1}}{\tau^2} = \frac{1}{2} A(v^{n+1} + v^{n-1}) + F^n, \\ v^0 = 0, \quad v^1 = \frac{1}{2}\tau^2 (E - \frac{\tau^2}{2} A)^{-1} F^0, \end{array} \right. \tag{2.53}$$

where A is the difference ahalogue of the operator $-\frac{\partial^2}{\partial x^2} + \lambda^2$,

$$F^n = F(v^n, \Phi_0^n, \hat{f}^n), \quad \Phi_0 = \varphi_{0tt} - \varphi_{0xx},$$

$$F^n = \frac{\hat{f}^n}{f_0^n} \left\{ \Phi_0^n + \lambda_n \left[R^2 \frac{v_{i1}^n + v_{iN+1}^n}{2} + \sum_{i=2}^{N} (-R - (j-1)\lambda_R)^2 v_{ij}^n \right] \right\}.$$

Lemma 2.4. Scheme (2.79) is absolutely stable and approximates problem (2.77) with the order $O(\tau^2 + h^2)$.

Together with (2.79) we have

$$\begin{aligned} h_i^{n+1} &= \frac{1}{f_0^{n+1}} \left\{ \Phi_{0i}^{n+1} + \lambda_R \left[R^2 \frac{v_{i1}^{n+1} + v_{iN+1}^{n+1}}{2} \right.\right. \\ &\qquad \left.\left. + \sum_{i=2}^{N} (-R + (j-1)\lambda_R)^2 v_{ij}^{n+1} \right] \right\}, \\ \frac{u^{n+1} - 2u^n + u^{n-1}}{\tau^2} &= \frac{1}{2} A_1 (u^{n+1} + u^{n-1}) + f^{n+1} h^{n+1}, \\ u^0 = 0, \quad u^1 &= \frac{1}{2}\tau^2 \left(E - \frac{\tau^2}{2} A_1 \right)^{-1} f^1 h^1, \end{aligned} \tag{2.54}$$

where A_1 is the difference analog of $\Delta = \frac{\partial^2}{\partial x^2} + \frac{\partial^2}{\partial z^2}$. Scheme (2.80) is also absolutely stable and approximates (2.71), (2.72) with the order $O(\tau^2 + h_1^2 + h_2^2)$.

The algorithm *A-II* being realized on FORTRAN consists in the following.

First step: in the domain

$$Q_{0\tau,h_1} = \{(0,\pi)\times(0,T)\}_{\tau,h_1} \ni (x,t)_{\tau h_1}$$

we solve problem (2.74).

Second step. We find $h(x,t)$ by the formula

$$h(x,t) = \frac{1}{f_0(x,t)}\left[\varphi_{0tt} - \varphi_{0xx} + \int_{-R}^{R} \lambda^2 v^R dx\right], \quad (x,t)_{\tau h_1} \in Q_{0\tau h_1}.$$

Third step. We recover the wave field

$$u(x,z,t) = \int_{-R}^{R} v^R(x,t,\lambda) e^{-i\lambda z} d\lambda,$$

$$u(x,z,t)_{\tau h_1 h_2} \in \{(0,\pi)\times\mathcal{D}\times(0,T)\}_{\tau h_1 h_2} \equiv Q_{\tau h_1 h_2}.$$

Inverse problems that arise in "shallow water" theory. Using linear shallow water theory (Whizem, 1977), we shall consider the problems of simulation of one-dimensional waves that depend on movement along the container bottom.

The mathematical setting is as follows:

$$\frac{\partial}{\partial t}(\eta + h) + \frac{\partial}{\partial x}[u(\eta + h)] = 0, \tag{2.55}$$

$$\frac{\partial u}{\partial t} + u\frac{\partial u}{\partial x} = g\frac{\partial \eta}{\partial x}, \tag{2.56}$$

where $u(x,t)$ is the horizontal component of velocity; $\eta + h_0$ is the full depth from the surface $y = \eta$ to the bottom of $y = -h$ (Fig. 2.1); $h(x,t) = h_0(x,t) - \eta_0(x,t)$ is the depth from a nonperturbed surface to the bottom $y = -h$; $\eta_0(x,t)$ is the local perturbation of the bottom.

Linearizing (2.81) and (2.82), we obtain the following system

$$u_t + g\eta_x = 0, \quad \eta_t + (uh_0)_x = \eta_{0t}, \quad (h_0 = h_0(x)),$$

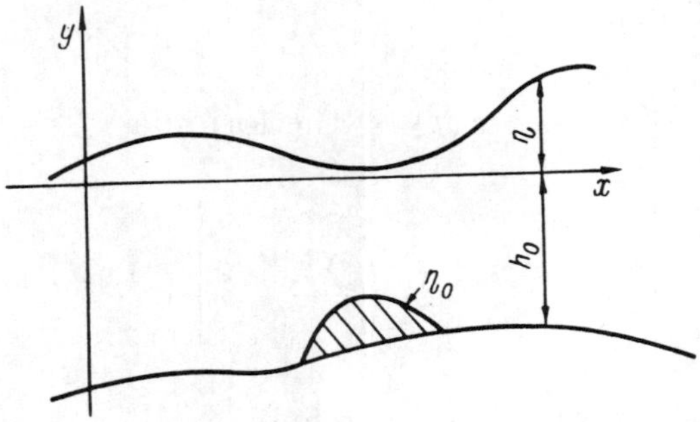

Figure 2.1. The wave creation process.

or, excluding u, we have

$$\eta_{tt} - g(\eta_x h_0)_x = \eta_{0tt}. \tag{2.57}$$

Further, the local bottom perturbation was taken as $\eta_0(x,t) = \theta(t)\alpha(x)$. Depending on the function $\alpha(x)$ and $\eta_0(x)$, the Cauchy problem for equation (2.83) was solved.

One of the inverse problems for (2.83) consists of determining functions $\alpha(x)$ and $\eta(x,t)$ by the measured profile of the wave $\eta(x,t)$ in the moment $T = t$. We may consider other inverse problems which consist in determining the bottom movement η_0 and the surface η.

Consider the two-dimensional shallow water theory. The equations of motion and conservation are as follows

$$u_{it} + u_j u_{ix_j} = -g\eta_{x_i}, \quad i = 1,2; \quad j = 1,2;$$

$$(h+\eta)_t + ((h+\eta)u_1)_{x_1} + ((\eta + h)u_2)_{x_2} = 0, \tag{2.58}$$

where $h + \eta$ is the full depth from the surface $y = \eta$ to the bottom $y = -h_0$; u_1, u_2 are the horizontal components of velocity; $h = h_0(x_1, x_2) - \eta_0(x_1, x_2, t)$ is the depth from the unperturbed surface to the bottom $y = -h$; $\eta_0(x_1, x_2, t)$ is the local perturbation of the bottom.

After linearization in (2.84) we obtain the system

$$u_{1t} = -g\eta_{x_1}, \quad u_{zt} = -g\eta_{x_2},$$

$$\eta_t + (h_0 u_1)_{x_1} + (h_0 u_2)_{x_2} = \eta_{0t}.$$

Excluding u_1, u_2, we obtain the equation for η

$$\eta_{tt} - g(h_0 \eta_{x_1})_{x_1} - g(h_0 \eta_{x_2})_{x_2} = \eta_{0tt}. \tag{2.59}$$

Adjoint to (2.85) the initial data in the moment $t = 0$ are

$$\eta\Big|_{t=0} = \eta_t\Big|_{t=0} = 0, \tag{2.60}$$

which are obtained on the basis of system (2.84) and the assumption that in the initial moment the medium was unperturbed.

Take $\eta_0(t, x_1, x_2)$ in the form

$$\eta_{0tt}(t, x_1, x_2) \equiv f(t)\alpha(x_1, x_2), \tag{2.61}$$

where $f(t)$ is a finite function, $\mathrm{supp} f \in (0, T)$.

Inverse problem. To find functions η, α (with f given) entering in (2.85), satisfying (2.86) and

$$\eta\Big|_{t=T_0} = \tau(x_1, x_2). \tag{2.62}$$

For the problem (2.85), (2.86), (2.88), two cases are possible.

1. $0 < T_0 < T$. Denote

$$\tilde{\eta} = \begin{cases} \eta, & t \geq 0, \\ 0, & t \leq 0, \end{cases} \qquad \tilde{f}(t) = \begin{cases} f, & t \in (0,T), \\ 0, & t \notin (0,T). \end{cases}$$

Then inverse problem (2.85), (2.86), (2.88) is equivalent to the following problem

$$p^2 v - g(h_0 v_{x_1})_{x_1} - g(h_0 v_{x_2})_{x_2} = \frac{\hat{f}(p)}{f(T_0)} \left[\int\limits_{-\infty}^{+\infty} p^2 v e^{pT_0} dp_0 - \Phi \right],$$

$$(x_1, x_2) \in R^2,$$

$$p = p_1 + ip_2, \;\; p_1 \geq p_0 > 0, \;\; \Phi = g(h_0 \tau_{x_1})_{x_1} + g(h_0 \tau_{x_2})_{x_2}, \tag{2.63}$$

$$\hat{f}(p) = \frac{1}{2\pi} \int\limits_{-\infty}^{+\infty} \tilde{f}(t) e^{-p_1 t - ip_2 t} dt.$$

The solutions to the inverse problem are as follows

$$\tilde{\eta} = e^{p_1 t} \int\limits_{-\infty}^{+\infty} v(p, x_1, x_2) e^{ip_2 t} dp_2, \;\; \alpha = \frac{1}{f(T_0)} \left[\int\limits_{-\infty}^{+\infty} p^2 v e^{pT_0} dp_2 - \Phi \right].$$

In order to give the scheme of solvability to problem (2.89), we first obtain *a priori* estimates.

Formulas (2.89) yield

$$\int\limits_{-\infty}^{+\infty} \int\limits_{R^2} |p|^{2m+2} |v|^2 dx_1 dx_2 dp_2 + \int\limits_{-\infty}^{+\infty} \int\limits_{R^2} g h_0 |\nabla v|^2 |p|^{2m} dx_1 dx_2 dp_2$$

$$\leq \int\limits_{-\infty}^{+\infty} \int\limits_{R^2} \frac{|v||\hat{f}|}{|f(T_0)|} |p|^{2m} \left[|\Phi| + e^{p_1 T_0} \int\limits_{-\infty}^{+\infty} |p|^2 |v| dp_2 \right] dx_1 dx_2 dp_2$$

$$+ \int\limits_{-\infty}^{+\infty} \int\limits_{R^2} \frac{|p_2||p|^{2m}}{p_1 |f(T_0)|} |v||\hat{f}| \left[|\Phi| + e^{p_1 T_0} \int\limits_{-\infty}^{+\infty} |p|^2 |v| dp_2 \right] dx_1 dx_2 dp_2. \tag{2.64}$$

In (2.90), the number $m \geq 3$ and $f(t) \in \overset{0}{W}{}_2^{m+1}(0,T)$; therefore, $\tilde{f}(t) \in W_2^m(R)$.

Estimating the integrals from the right side of (2.90), we obtain

$$\int_{-\infty}^{+\infty}\int_{R^2} \left|p\right|^{2m+2}\left|v\right|^2 dx_1 dx_2 dp_2 + \int_{-\infty}^{+\infty}\int_{R^2} gh_0 \left|p\right|^{2m}\left|\nabla v\right|^2 dx_1 dx_2 dp_2$$

$$\leq \frac{4e^{p_1 T_0}}{\left|f(T_0)\right|} p_1^{\frac{1-2m}{2}}$$

$$\times \left(\int_{-\infty}^{+\infty} \left|\hat{f}\right|^2 \left|p\right|^{2m} dp_2\right)^{1/2} \int_{-\infty}^{+\infty}\int_{R^2} \left|p\right|^{2m+2}\left|v\right|^2 dx_1 dx_2 dp_2$$

$$+\frac{\varepsilon}{2}\left(1+\frac{1}{p_1^2}\right)\int_{-\infty}^{+\infty}\int_{R^2} \left|p\right|^{2m+2}\left|v\right|^2 dx_1 dx_2 dp_2$$

$$+\frac{1}{2\varepsilon}\int_{-\infty}^{+\infty}\int_{R^2} \frac{|\hat{f}|^2\Phi^2|p|^{2m}}{p_1^2|f(T_0)|^2} dx_1 dx_2 dp_2 + \frac{1}{2\varepsilon}\int_{-\infty}^{+\infty}\int_{R^2} \frac{|\hat{f}|^2\Phi^2|p|^{2m}}{|f(T_0)|^2} dx_1 dx_2 dp_2, \tag{2.65}$$

where ε is a positive arbitrary number.

To obtain this inequality we have used the inequalities

$$\int_{-\infty}^{+\infty}\int_{R^2} \left|p_2\right|\left|p\right|^{2m}\left|v\right|\left|\hat{f}\right|\left(\int_{-\infty}^{+\infty}\left|p\right|^2\left|v\right| dp_2\right) dx_1 dx_2 dp_2$$

$$\leq \int_{R^2}\left(\int_{-\infty}^{+\infty}\left|p_2\right|^2\left|p\right|^{2m}\left|v\right|^2 dp_2\right)^{1/2}\left(\int_{-\infty}^{+\infty}\left|p\right|^{2m}\left|\hat{f}\right|^2 dp_2\right)^{1/2}$$

$$\times\left(\int_{-\infty}^{+\infty}\left|p_2\right|^2\left|p\right|^{2m}\left|v\right|^2 dp_2\right)^{1/2}\left(\int_{-\infty}^{+\infty}\frac{dp_2}{|p|^{2m-2}}\right)^{1/2} dx_1 dx_2,$$

$$\left|p\right|^{2m} \leq \frac{1}{p_1^2}\left|p\right|^{2m+2}.$$

Inequality (2.91) together with the inequality

$$1 - \frac{4e^{p_1 T_0}}{\left|f(T_0)\right|} p_1^{\frac{3-2m}{2}} \left(\int_{-\infty}^{+\infty} \left|\hat{f}\right|^2 \left|p\right|^{2m} dp_2 \right)^{1/2} \geq \delta > 0 \qquad (2.66)$$

yield the estimate

$$\int_{-\infty}^{+\infty} \int_{R^2} \left|p\right|^{2m+2} \left|v\right|^2 dx_1 dx_2 dp_2 + \int_{-\infty}^{+\infty} \int_{R^2} g h_0 \left|p\right|^{2m} \left|\nabla v\right|^2 dx_1 dx_2 dp_2 \leq c(f, \Phi).$$

This estimate allows us to estimate the higher derivatives of v with respect to x_1, x_2. Further, we may establish solvability by the method of extending by parameter.

Note, that the class of functions satisfying (2.92) is nonempty. Indeed, from the Parseval inequality it follows that

$$\int_0^T |D_t^m f|^2 e^{-2p_1 t} dt = \int_{-\infty}^{+\infty} |\hat{f}|^2 |p|^{2m} dp_2.$$

As the function $f(t)$ we may take a function $f \in C_0^\infty[0, T]$ such that

$$f(t) \equiv 1, \quad t \in [T_1, T_2], \quad 0 < T_1 < T_0 < T_2 < T,$$

$$0 \leq f \leq 1, \quad t \in [0, T_1] \bigcup [T_2, T],$$

$$f(t) \equiv 0, \quad t \notin [0, T], \quad |D_t^m f| \leq \gamma_m T^{-m}.$$

Thus, taking a suitable T, we may ensure that (2.92) holds.

Remark. The inverse problem under the condition $h_0(x_1, \psi(x_1)) = 0$ is of interest. This condition denotes that the curve $x_2 = \psi(x_1)$ is a shore.

2. $T_0 \geq T$

This case may be reduced to the case considered in the second subparagraph of this chapter, where the necessary and sufficient solvability conditions were obtained and the kernel of solutions to the inverse problem was completely described.

Inverse problems for the Lamé equations system. In $Q_T = (0,T) \times R^3$ consider the problem

$$(x')\underline{u}_{tt} = \mu(x')\Delta\underline{u} + [\lambda(x') + \mu(x')]\,\nabla\mathrm{div}\underline{u} + \underline{F}(x,t), \tag{2.67}$$

$$\underline{u}\Big|_{t=0} = \underline{u}_t\Big|_{t=0} = 0, \tag{2.68}$$

where

$$x = (x', z),\quad x' \in R^2,\quad \Delta = \frac{\partial^2}{\partial x_1^2} + \frac{\partial^2}{\partial x_2^2} + \frac{\partial^2}{\partial z^2} = \Delta_{x_1} + \frac{\partial^2}{\partial z^2}.$$

Let

$$\underline{F}(x,t) = (f_1(x,t)h_1(x',t),\ \ f_2(x,t)h_2(x',t),\ \ f_3(x,t)h_3(x',t)).$$

Inverse problem. To find vector functions $\underline{u} = (u_1, u_2, u_3)$, $h = (h_1, h_2, h_3)$ ($\underline{f}$, a given vector function) satisfying (2.93), (2.94) and such that

$$\underline{u}\Big|_{z=0} = \underline{u}_0(x', t). \tag{2.69}$$

A. If $|\underline{f}(x', z, 0)| \geq \delta > 0$, then inverse problem (2.93)-(2.95) is equivalent to the Cauchy problem for the following system

$$p(x')v_{1tt} = \mu(x')\Delta'_x v_1 - \sigma^2\mu(x'_1)v_1 + [\lambda(x') + \mu(x')]\frac{\partial}{\partial x_1}[v_{1x_1} + v_{2x_2}]$$

$$+i\sigma\,[\lambda(x') + \mu(x')]\,v_{3x_1} + \frac{\hat{f}_1(x',t,\sigma)}{f_1(x',t,0)}\Big[\Phi_1(x',t)$$

$$+\mu(x')\int_{-\infty}^{+\infty}\sigma^2 v_1 d\sigma - i(\lambda+\mu)\int_{-\infty}^{+\infty}\sigma v_{3x_1} d\sigma\Big],$$

$$p(x')v_{2tt} = \mu(x')\Delta'_x v_2 - \sigma^2\mu(x')v_2 + [\lambda(x') + \mu(x')]\frac{\partial}{\partial x_2}(v_{1x_1} + v_{2x_2})$$

$$+i\sigma\left[\lambda(x')+\mu(x')\right]v_{3x_2}+\frac{\hat f_2(x',t,\sigma)}{f_2(x',t,0)}\Big[\Phi_2(x',t)$$

$$+\mu\int_{-\infty}^{+\infty}\sigma^2 v_2 d\sigma-i(\lambda+\mu)\int_{-\infty}^{+\infty}\sigma v_{3x_2}d\sigma\Big], \tag{2.70}$$

$$p(x')v_{3tt}=\mu(x')\Delta_{x'}v_3-\sigma^2\mu(x')v_3+\left[\lambda(x')+\mu(x')\right]\left[i\sigma v_{1x_1}\right.$$

$$\left.+i\sigma v_{2x_1}-\sigma^2 v_3\right]+\frac{\hat f_3(x',t,\sigma)}{f_3(x',t,0)}\Big[\Phi_3(x',t)$$

$$+(\lambda+2\mu)\int_{-\infty}^{+\infty}\sigma^2 v_3 d\sigma-(\lambda+\mu)\int_{-\infty}^{+\infty}i\sigma(v_{1x_1}+v_{2x_2})d\sigma\Big],$$

$$v_j\Big|_{t=0}=v_{jt}\Big|_{t=0}=0,\quad j=\overline{1,3};$$

$$\hat f_k(x',t,\sigma)=\frac{1}{2\pi}\int_{-\infty}^{+\infty}f_k(x',t,z)e^{i\sigma z}dz,\quad k=\overline{1,3};$$

$$\Phi_k(x',t)=p(x')u_{0ktt}-\mu(x')\Delta_{x'}u_{0k}-(\lambda+\mu)\frac{\partial}{\partial x_k}(u_{01x_1}+u_{02x_2}),$$

$$k=1,2;$$

$$\Phi_3=p(x')u_{03tt}-\mu(x')\Delta_{x'}u_{03},\quad u_0=(u_{01},u_{02},u_{03}).$$

After determining the solution to (2.96), we find the inverse problem solution by the formulas

$$u_k=\int_{-\infty}^{+\infty}v_k e^{-i\sigma z}d\sigma,\quad k=\overline{1,3};$$

$$h_k=f_k^{-1}(x',t,0)\Big[\Phi_k+\mu\int_{-\infty}^{+\infty}\sigma^2 v_k d\sigma$$

$$-i(\lambda+\mu)\int_{-\infty}^{+\infty}\sigma v_{3x_k}d\sigma\Big],\quad k=1,2;$$

$$h_3=f_3^{-1}(x',t,0)\Big[\Phi_3+(\lambda+2\mu)\int_{-\infty}^{+\infty}\sigma^2 v_3 d\sigma$$

$$-\,i(\lambda+\mu)\int_{-\infty}^{+\infty}\sigma(v_{1x_1}+v_{2x_2})d\sigma\Big]. \tag{2.71}$$

Remark. Inverse problem (2.93)-(2.94), or, system (2.96) was considered by Kemelkov (1989). In particular, for the case ρ, λ, μ = const, the difference scheme which approximates (2.96) was constructed and investigated.

B. ρ, λ, $\mu = \text{const}, 0 \notin \operatorname{supp} f_k(z)$.

Denote

$$\hat{f}_k(\xi_3)=\frac{1}{2\pi}\int_{-\infty}^{+\infty}f_k(z)e^{i\xi_3 z}dz,$$

$$\hat{h}_k(p,\xi')=\frac{1}{(2\pi)^2}\int_{R^2}\int_0^{\infty}h_k(x',t)e^{-pt}e^{i(\xi',x')}dtdx',\quad k=\overline{1,3},$$

$$p=p_1+ip_2,\quad p_1>0,$$

$$\hat{u}_{0k}(p,\xi')=\frac{1}{(2\pi)^2}\int_{R^2}\int_0^{\infty}u_{0k}(x',t)e^{-pt}e^{i(\xi',x')}dtdx', \tag{2.72}$$

$$a_j=\rho p^2+\mu|\xi|^2+(\lambda+\mu)\xi_j^2,\quad |\xi|^2=\xi_1^2+\xi_2^2+\xi_3^2,\quad j=\overline{1,3},$$

$$b_1=(\lambda+\mu)\xi_1\xi_2=b_2,\quad b_3=(\lambda+\mu)\xi_1\xi_2,$$

$$\gamma_1=(\lambda+\mu)\xi_1\xi_3;\quad \gamma_2=\gamma_3=(\lambda+\mu)\xi_2\xi_3,$$

$$I(p,\xi)=\frac{(\hat{f}_1\hat{h}_1\gamma_2-\hat{f}_2\hat{h}_2\gamma_2)(b_2a_3-b_3\gamma_2)-(\hat{f}_2\hat{h}_3a_3-\hat{f}_3\hat{h}_3\gamma_2)(a_1\gamma_2-b_2\gamma_1)}{(b_1\gamma_2-a_2\gamma_1)(b_2a_3-b_3\gamma_2)-(a_3a_2-\gamma_3\gamma_2)(a_1\gamma_2-b_2\gamma_1)}.$$

Then the solution to inverse problem (2.93)-(2.94) is determined from the following relations

$$\int_{-\infty}^{+\infty} I d\xi_3 = \hat{u}_{02}, \tag{2.73}$$

$$\int_{-\infty}^{+\infty} \frac{\hat{f}_1\hat{h}_1\gamma_2 - \hat{f}_2\hat{h}_2\gamma_1}{a_1\gamma_2 - b_2\gamma_1} d\xi_3 - \int_{-\infty}^{+\infty} \frac{b_1\gamma_2 - a_2\gamma_1}{a_1\gamma_2 - b_2\gamma_1} I d\xi_3 = \hat{u}_{01},$$

$$\int_{-\infty}^{+\infty}\left[\frac{\hat{f}_3\hat{h}_3}{a_3} - \frac{\gamma_3}{a_3} I\right] d\xi_3 - \int_{-\infty}^{+\infty} \frac{b_3}{a_3}\left[\frac{\hat{f}_1\hat{h}_1\gamma_2 - \hat{f}_2\hat{h}_2\gamma_1}{a_1\gamma_2 - b_2\gamma_1} - \frac{b_1\gamma_2 - a_2\gamma_1}{a_1\gamma_2 - b_2\gamma_1} I\right] d\xi_3 = \hat{u}_{03}.$$

Note, that system (2.99) is a linear algebraic system relative to $\hat{h}_k$. The problem on its solvability is open.

After determining $\hat{h}_k$ from (2.99), we determine the functions u_k by the formulas

$$u_2 = \frac{1}{2\pi i} \int_{p_1 - i\infty}^{p_1 + i\infty} \int_{R^3} I(p,\xi) e^{pt} e^{-i(\xi,x)} d\xi dp,$$

$$u_1 = \frac{1}{2\pi i} \int_{p_1 - i\infty}^{p_1 + i\infty} \int_{R^3} \left[\frac{\hat{f}_1\hat{h}_1\gamma_2 - \hat{f}_2\hat{h}_2\gamma_1}{a_1\gamma_2 - b_2\gamma_1} - \frac{b_1\gamma_2 - a_2\gamma_1}{a_1\gamma_2 - b_2\gamma_1} I\right] e^{pt} e^{-i(\xi,x)} d\xi dp,$$

$$u_3 = \frac{1}{2\pi i} \int_{p_1 - i\infty}^{p_1 + i\infty} \int_{R^3} \left[\frac{\hat{f}_3\hat{h}_3}{a_3} - \frac{\gamma_3}{a_3} I(p,\xi) - \frac{b_3}{a_3}\left(\frac{\hat{f}_1\hat{h}_1\gamma_2 - \hat{f}_2\hat{h}_2\gamma_1}{a_1\gamma_2 - b_2\gamma_1}\right.\right.$$
$$\left.\left. - \frac{b_1\gamma_2 - a_2\gamma_1}{a_1\gamma_2 - b_2\gamma_1} I(p,\xi)\right)\right] e^{pt - i(\xi,x)} d\xi dp.$$

§2.3 Control problems and inverse problems.

In this paragraph, some problems of control through the nature of the source are interpreted as inverse problems for differential equations.

The control considered here is described by two coefficients entering into the evolutionary equations and not depending on time. Similar problems where considered by Lions (1973) and Butkovskii (1985). If control data are given, the inverse problems are reduced to the boundary-value problems for nonlinear integro-differential equations or to boundary-value problems for infinite systems of nonlinear differential equations. We shall show the methods for their investigation.

The basic mathematical difficulties are connected with ill-posedness study of such settings, especially, the problem of solution existence. After solving these problems, the control functions are obtained constructively.

Using a portion of the control data, first, we write out nonlinear integro-differential equations. Then, on the basis of the obtained equation and taking into account the boundary-value information, we shall formulate the results of the existence, uniqueness and stability investigations.

Consider the evolutionary equation

$$u_t = Au + f(x,t), \quad x \in \mathcal{D}, \quad t \in R, \tag{3.1}$$

where A is a linear operator, independent of t; $\mathcal{D} \in R^n$ is a domain with a smooth boundary $\Gamma \in C^{2m}$.

Unknown coefficients, for example, $c(x)$ and $h(x)$, may enter into A or into $f(x,t)$:

$$A = A_0 + c(x), \quad f = h(x)g(x,t),$$

where an operator A_0 and a function $g(x,t)$ are given. So, $c(x)$, $h(x)$ are unknown control functions. These problems for which we have to determine solution $u(x,t)$ and coefficients $c(x)$, $h(x)$ entering into (2.100), if certain information on $u(x,t)$ is given, are usually called inverse problems (Lavrent'ev *et al.*, 1980).

Together with (2.100), the control data $u_0(x)$, $u_T(x)$ are given:

$$u\Big|_{t=0} = u_0(x), \quad u\Big|_{t=T} = u_T(x). \tag{3.2}$$

We shall also assume that solutions $u(x,t)$ and $f(x,t)$ have direct and inverse Fourier transforms by t and differentiating under the integral sign is correct

$$w(x,\lambda) = \frac{1}{2\pi} \int\limits_{-\infty}^{+\infty} u(x,t)e^{i\lambda t}dt, \quad f(x,\lambda) = \frac{1}{2\pi} \int\limits_{-\infty}^{+\infty} f(x,t)e^{i\lambda t}dt. \tag{3.3}$$

We shall obtain from (2.100)-(2.102) new integro-differential equations, connected with control problems by two certain coefficients entering into (2.100) when $u(x,t)$ passes from the state $u_0(x) = u\Big|_{t=0}$ into the state $u_T(x) = u\Big|_{t=T}$. For example, such coefficients are $c(x)$ and $h(x)$ for $A = A_0 + c(x)$, $f = h(x)g(x,t)$. These equations relative to $w(x,t)$ and two arbitrary coefficients are as follows

$$-i\lambda w = Aw + \hat{f}, \tag{3.4}$$

$$-\int\limits_{-\infty}^{+\infty} -i\lambda w d\lambda = Av_0 + f_0, \quad f_0 = f\Big|_{t=0},$$

$$-\int\limits_{-\infty}^{+\infty} -i\lambda e^{-i\lambda T} w d\lambda = Au_T + f_T, \quad f_T = f\Big|_{t=T}.$$

Analogously, for the equation $u_{tt} = Au + f$ with the control data $u_0 = u\Big|_{t=0}$, $u_T = u\Big|_{t=T}$, the following system is obtained

$$-\lambda^2 w = Aw + \hat{f},$$

$$-\int\limits_{-\infty}^{+\infty} \lambda^2 w d\lambda = Av_0 + f_0, \tag{3.5}$$

$$-\int\limits_{-\infty}^{+\infty} \lambda^2 w e^{-i\lambda T} d\lambda = Au_T + f_T.$$

This system, as before, we may consider as a system relative to $w(x,\lambda)$ and any two coefficients entering into A and f and responsible for control when $u(x,t)$ from the state $u_0(x)$ passes to the state $u_T(x)$.

If, instead of (2.102), we assume that $u(x,t)$ and $f(x,t)$ are periodic:

$$u = \sum_{k=-\infty}^{\infty} e^{ikt} u_k(x), \quad f = \sum_{k=-\infty}^{\infty} f_k(x) e^{ikt},$$

then systems (2.103) and (2.104) become infinite systems also relative to certain coefficients entering into A and f. These systems are the following

$$\begin{cases} iku_k = Au_k + f_k, \\ \sum_k iku_k = Au_0 + f_0, \\ \sum_k ike^{ikt} u_k = Au_T + f_T; \end{cases} \tag{3.6}$$

$$\begin{cases} -k^2 u_k = Au_k + f_k, \\ -\sum_k k^2 u_k = Au_0 + f_0, \\ -\sum_k k^2 u_k e^{ikt} = Au_T + f_T; \quad k = 0, 1, \dots . \end{cases} \tag{3.7}$$

Sometimes the desired coefficients may be excluded from systems (2.103)-(2.106), and the obtained equations will be only relative to $w(x, \lambda)$ or u_k.

Suppose that in (2.103), $A = A_0 + c(x)$, $f = h(x)g(x, t)$, where $h(x)$ and $c(x)$ are the desired control functions; let A_0 be a linear operator independent of t; g be a given function.

If

$$|M| = |g_T u_0 - u_T g_0| \geq \sigma > 0, \quad g_T = g\Big|_{t=\pi}, \quad g_0 = g\Big|_{t=0},$$

then (2.103) yields

$$h(x) = \frac{1}{M}\Big[u_T \left(\int_{-\infty}^{+\infty} i\lambda w d\lambda + A_0 u_0 \right) - u_0 \left(\int_{-\infty}^{+\infty} i\lambda e^{-i\lambda T} w d\lambda + A_0 u_T \right)\Big], \tag{3.8}$$

$$c(x) = \frac{1}{M}\Big[g_0 \left(\int_{-\infty}^{+\infty} i\lambda e^{-i\lambda T} w d\lambda + A_0 u_T \right)$$

$$- g_T \left(\int_{-\infty}^{+\infty} i\lambda w d\lambda + A_0 u_0 \right) \Big]. \tag{3.9}$$

Taking into account (2.107), (2.108), (2.103), we obtain the nonlinear integro-differential equation relative to $w(x, \lambda)$

$$\begin{aligned} A_0 w + i\lambda w + \frac{w}{M} \Big[g_0 \left(\int_{-\infty}^{+\infty} i\lambda e^{-i\lambda T} w d\lambda + A_0 u_T \right) \\ - g_T \left(\int_{-\infty}^{+\infty} i\lambda w d\lambda + A_0 u_0 \right) \Big] \\ + \frac{\hat{g}(x, \lambda)}{M} \Big[u_T \left(\int_{-\infty}^{+\infty} i\lambda w d\lambda + A_0 u_0 \right) \\ - u_0 \left(\int_{-\infty}^{+\infty} i\lambda e^{-i\lambda T} w d\lambda + A_0 u_T \right) \Big] = 0. \end{aligned} \tag{3.10}$$

For the obtained systems (2.103), (2.106), (2.109), if we want to select a concrete solution, we need to add the boundary conditions. We shall restrict ourselves to equation (2.109) and, moreover, assume the operator A_0 to be an elliptic operator of order $2m$, with coefficients of the class $C^{2m}(0)$, and for which the Dirichlet problem is uniquely solvable.

So, consider equation (2.109) with boundary conditions

$$w\Big|_\Gamma = \hat{\varphi}_0, \quad \frac{\partial w}{\partial n} = \hat{\varphi}_1, \dots, \frac{\partial^{m-1} w}{\partial n^{m-1}}\Big|_\Gamma = \hat{\varphi}_{m-1}, \tag{3.11}$$

where

$$\hat{\varphi}_k(s, \lambda) = \frac{1}{2\pi} \int_{-\infty}^{+\infty} \frac{\partial^k}{\partial \eta^k} u\Big|_\Gamma e^{i\lambda t} dt.$$

After determining the solution $w(x, \lambda)$ to problem (2.109), (2.110) from the corresponding class, the control functions $h(x)$, $c(x)$ are determined by formulas (2.107), (2.108).

Thus, in this case, the basic problem is to investigate the well-posedness of problem (2.109), (2.110). In order to illustrate the possibilities arising here, we shall formulate below a theorem on the unique solvability of this problem.

Theorem 2.9. Let $g, \varphi_0, \ldots, \varphi_{m-1}$ be such that

$$E^2(g, \varphi_0, \ldots, \varphi_{m-1}) \equiv \int_{-\infty}^{+\infty} \int_{\mathcal{D}} \left(1 + |\lambda|^{2+2p}\right) |\hat{g}| d\lambda d\mathcal{D}$$

$$+ \int_{-\infty}^{+\infty} \left(1 + |\lambda|^{4+2p}\right) \sum_{i=0}^{m-1} \left\|\hat{\varphi}_i\right\|^2_{W_2^{2m-i-1/2}(\Gamma)} d\lambda < \infty, \ \ 2p > 1$$

and, besides,

$$n \leq 3, \ \ u_0, u_T \in C^{2m}(\bar{\mathcal{D}}); \ \ g_0, g_t \in C(\bar{\mathcal{D}}),$$

$$M = C_0 \frac{T(g, \varphi_0, \ldots, \varphi_{m-1})}{(2p-1)^{1/2}} - \left\|\frac{\Phi_0}{\mu}\right\|_0 > 0,$$

where the constant C_0 depends on u_0, u_T, g_0, g_T; $\Phi_0 = -g_0 A_0 u_T + g_T A_0 u_0$; μ is a constant from the inequality

$$\left\|A_0 v\right\| \geq \mu \left\|v\right\|_{W_2^m}, \ \ \ v \in \overset{0}{W}{}_2^m(\mathcal{D}) \bigcap W_2^{2m}(\mathcal{D}) < \infty.$$

Then there exists a unique solution $w(x, \lambda)$ to problem (2.109), (2.110) and

$$\int_{-\infty}^{+\infty} \left(1 + |\lambda|\right)^{2p+2} \left\|w\right\|^2_{L^2(\mathcal{D})} d\lambda + \int_{-\infty}^{+\infty} \left(1 + |\lambda|\right)^{2p} \left\|w\right\|^2_{W_2^{2m}(\mathcal{D})} d\lambda < \infty.$$

§2.4 Inverse problems for nonlinear equations

In this paragraph we shall consider the inverse problems for systems of nonlinear equations of the form

$$\frac{\partial \bar{u}}{\partial t} = A\bar{u} + \bar{c}(\bar{u}), \qquad \frac{\partial^2 \bar{u}}{\partial t^2} = A\bar{u} + \bar{c}(\bar{u}), \tag{4.1}$$

where A is a given linear operator, for example, the Laplace operator. The vector functions $\bar{c}(\bar{u})$, $\bar{u}(\bar{x}, t)$ are unknown. The information is the trace of solution $\bar{u}(\bar{x}, t)$ and, further, we know some qualitative information. For example, the solution is periodic with respect to t or to a certain spatial variable. On the basis of this information we shall deduce new integro-differential equations, which do not contain the unknown functions $\bar{c}(\bar{u})$. Therefore, the inverse problems are reduced to initial boundary-value problems for the obtained equations. To solv them, then, we calculate the desired functions $\bar{c}(\bar{u})$ entering into (2.111). It seems that these problems may be interesting for future research. Here we shall only formulate the uniqueness of solution in the class of analytic functions.

Before we begin to deduce those equations not containing the sources $\bar{c}(\bar{u})$, note that the periodicity conditions for equations (2.111) have already been discussed (Marri, 1983; Smoes, 1984).

Here we shall assume that $\bar{x} \in \mathcal{D} \subset R^4$, $t \in R$, $\bar{u} \in R^n$, $\bar{c} \in R^n$, $\mathcal{D}$ is a domain with a smooth boundary. An operator A is linear and does not depend on t. Besides, for solutions $\bar{u}(x, t)$ to (2.111) the information

$$\bar{u}_0 = \bar{u}\Big|_{t=0}, \qquad \bar{x} \in \mathcal{D}, \tag{4.2}$$

$$\exists \bar{w}(x, \bar{\omega}) = \frac{1}{2\pi} \int_{-\infty}^{+\infty} \bar{u}(\bar{x}, t) e^{-i\omega t} dt \tag{4.3}$$

is given. Applying this information, we obtain the integro-differential equations relative to the vector function $\bar{w}(\bar{x}, t)$ which do not contain $\bar{c}(\bar{u})$. In the case

$$\frac{\partial \bar{u}}{\partial t} = A\bar{u} + \bar{c}\bar{u},$$

the equation is as follows

$$A\bar{w} - i\omega\bar{w} = St\bar{w}, \tag{4.4}$$

where

$$St\bar{w} = -\frac{1}{2\pi}\int\limits_{-\infty}^{+\infty}\left[\int\limits_{-\infty}^{+\infty} is\bar{w}\left(\bar{u}_0^{-1}\left(\int\limits_{-\infty}^{+\infty}\bar{w}e^{i\eta t}d\eta\right), s\right) ds\right.$$
$$\left. + A\bar{u}_0\left(\bar{u}_0^{-1}\left(\int\limits_{-\infty}^{+\infty}\bar{w}e^{i\eta t}d\eta\right)\right)\right] e^{-i\omega t}dt.$$

In the case

$$\frac{\partial^2\bar{u}}{\partial t^2} = A\bar{u} + \bar{c}\bar{u},$$

the equation for $\bar{w}(x,\omega)$ is as follows

$$A\bar{w} + \omega^2\bar{w} = St\bar{w},$$

where

$$St\bar{w} = -\frac{1}{2\pi}\int\limits_{-\infty}^{+\infty}\left[\int\limits_{-\infty}^{+\infty} -\xi^2\bar{w}\left(\bar{u}_0^{-1}\left(\int\limits_{-\infty}^{+\infty}\bar{w}e^{i\eta t}d\eta\right), \xi\right) d\xi\right.$$
$$\left. + A\bar{u}_0\left(\bar{u}_0^{-1}\left(\int\limits_{-\infty}^{+\infty}\bar{w}e^{i\eta t}d\eta\right)\right)\right] e^{-i\omega t}dt. \tag{4.5}$$

Here $\bar{x} = \bar{u}_0^{-1}(-p)$ is the inverse mapping to $\bar{u}_0(\bar{x})$.

We shall assume that (2.111), (2.112) hold, and (2.113) is changed through its periodicity with respect to t. Besides, we shall suppose that solutions $\bar{u}(\bar{x},t)$ to equation (2.111) are represented as the Fourier series

$$\bar{u}(\bar{x},t) = \sum_{k=0}^{\infty}\left(\bar{u}_k(x)\cos kt + \bar{v}_k(x)\sin kt\right). \tag{4.6}$$

Taking into account (2.111), (2.112), (2.116), we obtain relative to $\bar{u}_k(x)$, $\bar{v}_k(x)$, $k = 0,1,2,\ldots$ the nonlinear integro-differential infinite systems which do not contain the desired functions $\bar{c}(\bar{u})$. For

$$\frac{\partial \bar{u}}{\partial t} = A\bar{u} + \bar{c}(\bar{u})$$

these systems have the form

$$A\bar{u}_k - k\bar{v}_k = -\frac{1}{\pi}\int_{-\pi}^{\pi}\left[\sum_{j=1}^{\infty} j\bar{v}_j\left(\bar{u}_0^{-1}\left(\sum_{s=0}^{\infty}\bar{u}_s\cos st + \bar{v}_s\sin st\right)\right)\right.$$
$$\left. -A\bar{u}_0\left(\bar{u}_0^{-1}\left(\sum_{s=0}^{\infty}\bar{u}_s\cos st + \bar{v}_s\sin st\right)\right)\right]\cos kt dt,$$
$$A\bar{v}_k - k\bar{u}_k = -\frac{1}{\pi}\int_{-\pi}^{\pi}\left[\sum_{j=1}^{\infty} j\bar{v}_j\left(\bar{u}_0^{-1}\left(\sum_{s=0}^{\infty}\bar{u}_s\cos st + \bar{v}_s\sin st\right)\right)\right.$$
$$\left. -A\bar{u}_0\left(\bar{u}_0^{-1}\left(\sum_{s=0}^{\infty}\bar{u}_s\cos st + \bar{v}_s\sin st\right)\right)\right]\sin kt dt, \tag{4.7}$$

Analogously, in the case

$$\frac{\partial^2 \bar{u}}{\partial t^2} = A\bar{u} + \bar{c}(\bar{u}),$$

the system is as follows

$$A\bar{u}_k - k^2\bar{u}_k = \frac{1}{\pi}\int_{-\pi}^{\pi}\left[\sum_{j=1}^{\infty} j^2\bar{u}_j\left(\bar{u}_0^{-1}\left(\sum_{s=0}^{\infty}\bar{u}_s\cos st + \bar{v}_s\sin st\right)\right)\right.$$
$$\left. -A\bar{u}_0\left(\bar{u}_0^{-1}\left(\sum_{s=0}^{\infty}\bar{u}_s\cos st + \bar{v}_s\sin st\right)\right)\right]\cos kt dt,$$
$$A\bar{v}_k - k^2\bar{v}_k = \frac{1}{\pi}\int_{-\pi}^{\pi}\left[\sum_{j=1}^{\infty} j^2\bar{u}_j\left(\bar{u}_0^{-1}\left(\sum_{s=0}^{\infty}\bar{u}_s\cos st + \bar{v}_s\sin st\right)\right)\right.$$
$$\left. -A\bar{u}_0\left(\bar{u}_0^{-1}\left(\sum_{s=0}^{\infty}\bar{u}_s\cos st + \bar{v}_s\sin st\right)\right)\right]\sin kt dt, \tag{4.8}$$

Now, in the domain $\bar{x} \in \mathcal{D} \subset R^{n-1}$, $z \in R$, $T_0 \leq t \leq T$, consider the nonlinear system of evolutionary equations

$$\frac{\partial \bar{u}}{\partial t} = A\bar{u} + \frac{\partial^2 \bar{u}}{\partial z^2} + \bar{c}(\bar{u}), \tag{4.9}$$

where A is a linear operator, independent of z; for example,

$$A = \sum_{g=1}^{n-1} \frac{\partial^2}{\partial x_g^2}, \quad \bar{u} = (u_1, u_2, \ldots, u_n), \quad \bar{c} = (c_1, c_2, \ldots, c_n), \quad n \geq 1;$$

the spatial variable z is selected. Assume that the following two conditions hold: we know the solution (2.119) trace for $z = 0$, i.e.,

$$\bar{u}\Big|_{z=0} = \bar{u}_0(\bar{x}, t), \tag{4.10}$$

and the vector function $\bar{u}_0(\bar{x}, t)$ is given.

There exists the Fourier transformation of the vector function $\bar{u}(\bar{x}, z, t)$ by z

$$\bar{w}(\bar{x}, \omega t) = \frac{1}{2\pi} \int_{-\infty}^{+\infty} \bar{u}(\bar{x}, z, t) e^{-iz\omega} dz. \tag{4.11}$$

Let $\bar{u}_0^{-1}(g)$ be the inverse mapping to $\bar{u}_0(\bar{x}, t)$; $(\bar{x}, t) = \bar{p} \in R^n$. If these conditions (or other analogous conditions) hold, the excluding of $\bar{c}(\bar{u})$ from (2.111) leads to an essentially nonlinear evolutionary equation for the function $\bar{w}(\bar{x}, \omega, t)$. This equation is the following

$$\frac{\partial \bar{w}}{\partial t} = A\bar{w} + St\bar{w},$$

where

$$St\bar{w} = -\frac{\partial \bar{w}}{\partial t} - \omega^2 \bar{w} + \int_{-\infty}^{+\infty} \left\{ \frac{\partial \bar{u}_0}{\partial t} \left(\bar{u}_0^{-1} \left(\int_{-\infty}^{+\infty} \bar{w}(\bar{x}, y, t) e^{izy} dy \right) \right) \right\}$$

$$-A\bar{u}_0 \left(\bar{u}_0^{-1} \left(\int_{-\infty}^{+\infty} \bar{w}(\bar{x}, y, t) e^{izy} dy \right) \right)$$

$$+ \int_{-\infty}^{+\infty} \xi^2 \bar{w} \left(\bar{u}_0^{-1} \left(\int_{-\infty}^{+\infty} \bar{w}(\bar{x}, y, t) e^{izy} dy \right), \xi \right) e^{-iz\omega} dz. \tag{4.12}$$

Here, nonlinearity consists in dependence of known and unknown functions on the integral

$$\int_{-\infty}^{+\infty} \bar{w}(\bar{x},y,t)e^{izy}dy.$$

Analogously, for the system

$$\frac{\partial^2 u}{\partial t^2} = A\bar{u} + \frac{\partial^2 \bar{u}}{\partial z^2} + \bar{c}(\bar{u}), \tag{4.13}$$

with the same conditions (2.120), (2.121), we may write out the equation relative to $\bar{w}(\bar{x},\omega,t)$, which does not contain $\bar{c}(\bar{u})$

$$\frac{\partial^2 \bar{w}}{\partial t^2} = A\bar{w} + St\bar{w},$$

$$\frac{\partial^2 \bar{w}}{\partial t^2} = A\bar{w} - \omega^2\bar{w} + \int_{-\infty}^{+\infty} \left\{ \frac{\partial^2 \bar{u}_0}{\partial t^2} \left(\bar{u}_0^{-1} \left(\int_{-\infty}^{+\infty} \bar{w}(\bar{x},y,t)e^{izy}dy \right) \right) \right.$$

$$- A\bar{u}_0 \left(\bar{u}_0^{-1} \left(\int_{-\infty}^{+\infty} \bar{w}(\bar{x},y,t)e^{izy}dy \right) \right)$$

$$\left. + \int_{-\infty}^{+\infty} \xi^2 \bar{w} \left(\bar{u}_0^{-1} \left(\int_{-\infty}^{+\infty} \bar{w}(\bar{x},y,t)e^{izy}dy \right), \xi \right) d\xi \right\} e^{-iz\omega}dz. \tag{4.14}$$

If we have the initial boundary-value information for equations (2.119), (2.123) and after solving initial boundary-value problems for equations (2.122), (2.124), we determine the unknown coefficients $\bar{c}(\bar{u})$ and solution $\bar{u}(\bar{x},z,t)$ by the formulas *via* $\bar{w}(\bar{x},\omega,t)$.

Now consider the periodic case. Namely, instead of condition (2.121), which was assumed to hold for equations (2.119), (2.123), we shall suppose that $\bar{u}(\bar{x},z,t)$ is represented as the Fourier series in z

$$\bar{u}(\bar{x},z,t) = \sum_{k=0}^{\infty} \left(\bar{u}_k(\bar{x},t)\cos kz + \bar{v}_k(\bar{x},t)\sin kz \right). \tag{4.15}$$

Let $\bar{\Phi}(\bar{p}) = \partial\bar{u}_0/\partial t - A\bar{u}_0$, $\bar{p} = (\bar{x}, t)$. Using (2.119), (2.120), and (2.125) we have the equations relative to $\bar{u}_k(\bar{x}, t)$, $\bar{v}_k(\bar{x}, t)$ without $\bar{c}(\bar{u})$

$$\frac{\partial \bar{u}}{\partial t} = A\bar{u}_k - k^2\bar{u}_k + \frac{1}{2\pi}\int_{-\pi}^{\pi}\left[\Phi\left(\sum_{j=0}^{\infty}(\bar{u}_j \cos jz + \bar{v}_j \sin jz)\right)\right.$$

$$\left.+\sum_{s=1}^{\infty} s^2\bar{u}_s\left(\bar{u}_0^{-1}\left(\sum_{j=0}^{\infty}\bar{u}_j\cos jz + \bar{v}_j \sin jz\right)\right)\right]\cos kz dz;$$

$$\frac{\partial \bar{v}_k}{\partial t} = A\bar{v}_k - k^2\bar{v}_k + \frac{1}{2\pi}\int_{-\pi}^{\pi}\left[\Phi\left(\sum_{j=0}^{\infty}(\bar{u}_j \cos jz + \bar{v}_j \sin jz)\right)\right.$$

$$\left.+\sum_{s=1}^{\infty} s^2\bar{u}_s\left(\bar{u}_0^{-1}\left(\sum_{j=0}^{\infty}\bar{u}_j\cos jz + \bar{v}_j \sin jz\right)\right)\right]\sin kz dz. \tag{4.16}$$

Analogously, in the case

$$\frac{\partial^2 \bar{u}}{\partial t^2} = A\bar{u} + \bar{c}(\bar{u}),$$

the system relative to $\bar{u}_k(\bar{x}, t), \bar{v}_k(\bar{x}, t)$ is as follows

$$\frac{\partial^2 \bar{u}}{\partial t^2} = A\bar{u}_k - k^2\bar{u}_k + \frac{1}{2\pi}\int_{-\pi}^{\pi}\left[\bar{\psi}\left(\sum_{j=0}^{\infty}(\bar{u}_j \cos jz + \bar{v}_j \sin jz)\right)\right.$$

$$\left.+\sum_{s=1}^{\infty} s^2\bar{u}_s\left(\bar{u}_0^{-1}\left(\sum_{j=0}^{\infty}\bar{u}_j\cos jz + \bar{v}_j \sin jz\right)\right)\right]\cos kz dz;$$

$$\frac{\partial^2 \bar{v}_k}{\partial t^2} = A\bar{v}_k - k^2\bar{v}_k + \frac{1}{2\pi}\int_{-\pi}^{\pi}\left[\bar{\psi}\left(\sum_{j=0}^{\infty}(\bar{u}_j \cos jz + \bar{v}_j \sin jz)\right)\right.$$

$$\left.+\sum_{s=1}^{\infty} s^2\bar{u}_s\left(\bar{u}_0^{-1}\left(\sum_{j=0}^{\infty}\bar{u}_j\cos jz + \bar{v}_j \sin jz\right)\right)\right]\sin kz dz. \tag{4.17}$$

where $\bar{\psi}(\bar{p}) = \partial^2\bar{u}_0/\partial t^2 - A\bar{u}_0$, $\bar{p} = (\bar{x}, t)$.

These equations are essentially nonlinear. If we have additional information, i. e., information typical for inverse problems, initial boundary-value problems arise for these equations. After solving these problems, the desired vector functions $\bar{c}(\bar{u})$ are obtained by the formulas.

Untile now we may formulate the result of uniqueness of the solution only in the analytic case.

Let A be a differential operator with a higher derivative with respect to x_1 and assume that it is solved relative to this derivative: $\partial^m \bar{u}/\partial x_1^m$.

Assume also that we have the additional information

$$\left.\frac{\partial^k \bar{u}}{\partial x_1^k}\right|_{x_1=0}, \quad k = 0, 1, 2, \ldots, m-1.$$

Then, bearing in mind the character of equations (2.114), (2.115), (2.117), (2.118), (2.122), (2.124), (2.126), (2.127), we may for corresponding restrictions formulate the following result: in the case of analytic functions, the inverse problems have no more than one solution.

Existence problems and another boundary-value problems (therefore, other inverse problems) are not investigated.

The essential peculiarity of evolutionary processes is their nonlinearity (Marri, 1983; Leibovich and Sibass, 1977; Math. Modelling, 1986; Synergetics, 1984). Moreover, the evolutionary laws are often unknown and are to be found. In other words, parallel with solutions we have to find those evolutionary equations which these solutions satisfy. Such problems are characteristic of the theory and applications of multidimensional inverse problems for differential equations, in particular, for equations of mathematical physics (Alekseev, 1962; Anikonov, 1978, 1987; Lavrent'ev, 1972; Tikhonov and Arsenin, 1979).

In the case of nonlinear equations, the inverse problem subjects are interlacing with problems of synergetics (as the science of mathematical methods for investigating nonlinear phenomena and the properties of these phenomena). Here we shall give ways of excluding such indeterminacies of nonlinear evolutionary equations, for the most part, based on empirical data. The new equations and formulas, obtained as a result characterize the synergetic property which is inherent in nonlinear evolutionary processes. The results given below are, for the most part, of a rather formal character.

1. In the domain $\mathcal{D} \times [0,T]$, $\mathcal{D} \subset R^n$ consider the nonlinear evolutionary system

$$\frac{\partial \bar{u}}{\partial t} = A\bar{u} + \bar{\lambda}(\bar{u}), \tag{4.18}$$

where $\bar{u}(x,t) = (u_1, u_2, \ldots, u_n)$; $\bar{x} \in \mathcal{D}$, $0 \le t \le T$; A is a linear differential operator depending only on x (nor depending on t); $\bar{\lambda}(\bar{u})$ is a vector function, which may be interpreted as the source function.

We shall suppose in (2.128) vector functions $\bar{u}(\bar{x},t)$ and $\bar{\lambda}(\bar{u})$ to be unknown. Assume that

$$\bar{u}\Big|_{t=0} = \bar{a}(\bar{x}), \quad \frac{\partial u}{\partial t}\Big|_{t=0} = \bar{b}(\bar{x}), \quad \bar{x} \in \mathcal{D}. \tag{4.19}$$

It occurs that from (2.128), (2.129), to within the inverse mapping $\bar{a}^{-1}(\bar{p})$ to $\bar{p} = \bar{a}(\bar{x}) = \bar{u}\Big|_{t=0}$ for $\bar{u}(\bar{x},t)$ we may write the equation system determined only by (2.129) and not depending on $\bar{\lambda}(\bar{u})$. This fact is formulated below.

Theorem 2.10. The evolution of $\bar{u}(\bar{x},t)$ is described by the system

$$\frac{\partial \bar{u}}{\partial t} = A\bar{u} + \bar{b}\left(\bar{a}^{-1}(\bar{u})\right) - A\bar{a}\left(\bar{a}^{-1}(\bar{u})\right), \tag{4.20}$$

where

$$\bar{u}\Big|_{t=0} = \bar{a}(\bar{x}), \quad \frac{\partial \bar{u}}{\partial t}\Big|_{t=0} = \bar{b}(\bar{x}), \quad \bar{\lambda}(\bar{p}) = \bar{b}\left(\bar{a}^{-1}(\bar{p})\right) - A\bar{a}\left(\bar{a}^{-1}(\bar{p})\right).$$

If instead of (2.129), for $-\infty < t < \infty$ the information

$$u\Big|_{t=0} = \bar{a}(\bar{x}) \quad \exists \bar{w}(\bar{x},\omega) = \frac{1}{2\pi} \int_{-\infty}^{+\infty} \bar{u}(\bar{x},t) e^{-i\omega t} dt \tag{4.21}$$

is given, then for the desired function $w(\bar{x},\omega)$ the system may be written, not depending on $\bar{\lambda}(\bar{u})$.

Theorem 2.11. The system for $\bar{w}(\bar{x},\omega)$, $x \in \mathcal{D}$, $\omega \in R$ is as follows

$$\mathcal{A}\bar{w} - i\omega\bar{w} = St\bar{w}, \tag{4.22}$$

where

$$St\bar{w} = -\frac{1}{2\pi}\int_{-\infty}^{+\infty}\left[\int_{-\infty}^{+\infty} i\varepsilon\bar{w}\left(\bar{a}^{-1}\left(\int_{-\infty}^{+\infty}\bar{w}(\bar{x},y)e^{ity}dy,\varepsilon\right)de\right)\right.$$
$$\left.-\mathcal{A}\bar{a}\left(\bar{a}^{-1}\int_{-\infty}^{+\infty}\bar{w}(\bar{x},y)e^{ity}dy\right)\right]e^{-ity}dt.$$

The system (2.132) is essentially nonlinear: the solution $\bar{w}(\bar{x},\omega)$ depends on the solution functionals.

2. The general kinematic approach of wave theory consists in introducing the phase function $\theta(x,t)$, $x \in \mathcal{D}$, $0 \le t \le T$ of the wave vector $\bar{K}(x,t) = \mathrm{grad}_x\theta(x,t)$, the frequency $\omega = -\frac{\partial\theta}{\partial t}$, and the dispersible relation, which in the homogeneous case is such $\omega = w(\bar{k})$. According to this theory, the phase evolution and the wave vector $\bar{K}(x,t)$ evolution satisfy the equations

$$\frac{\partial\theta}{\partial t} = -w(\mathrm{grad}_x\theta)$$

$$\frac{\partial K_j}{\partial t} + \sum_{i=1}^{n}\frac{\partial w}{\partial K_i}\frac{\partial K_i}{\partial x_j} = 0. \tag{4.23}$$

If the form of dispersible dependence $w = w(\bar{k})$ is *a priori* unknown, then equations (2.133) are also unknown. We shall consider here two settings of inverse problems.

To find $\bar{K}(x,t)$ and $\bar{w}(\bar{k})$ if one of the following two conditions holds

1) $\bar{k}\Big|_{t=0} = \bar{a}(\bar{x}), \qquad \frac{\partial\bar{k}}{\partial t}\Big|_{t=0} = \bar{b}(\bar{x});$

2) $\bar{k}\Big|_{t=0} = \bar{a}(\bar{x}), \qquad \bar{k}\Big|_{t=T} = \bar{c}(x).$

Theorem 2.12. In the case 1) we have

$$\bar{k}(\bar{x},t) = \bar{a}\left(\bar{x} + A^{-1}\big(\bar{a}^{-1}(\bar{k})\big)\bar{b}\big(\bar{a}^{-1}(\bar{k})\cdot t\big)\right),$$

$$\text{grad } w = -A^{-1}\big(\bar{a}^{-1}(\bar{k})\big)\bar{b}\big(\bar{a}^{-1}(\bar{k})\big),$$

where $\bar{a}^{-1}(\bar{k})$ is inverse to $\bar{p} = \bar{a}(x)$; A^{-1} is inverse to $A = \partial u_i/\partial x_j$, $i, j = 1, 2, \ldots, n$.

Theorem 2.13. In the case 2) we have

$$\bar{k}(x,t) = \bar{a}\left(x - \frac{c^{-1}(\bar{k}) - \bar{a}^{-1}(\bar{k})}{T}t\right),$$

$$\text{grad } w = \frac{c^{-1}(\bar{k}) - \bar{a}^{-1}(\bar{k})}{T}.$$

As above, the ambiguity may be because of multivalence of mappings $\bar{a}(\bar{x}) = \bar{p}$, $\bar{c}(\bar{x}) = q$.

3. Here we shall consider the equations, the coefficients of which are nonlinear functionals from the solutions. The general form of such equations is as follows:

$$\big(\mathcal{A}_x + L_z\big)u + \int\limits_{R^m} F\big(x, z, u, u_x, u_z\big)dz \cdot \big(\mathcal{B}_x + \mathcal{D}_z\big)u = 0, \tag{4.24}$$

where $\mathcal{A}_x, \mathcal{B}_x$ are linear differential operators acting with respect to $x \in R^n$; L_z, D_z are linear differential operators acting with respect to $z \in R^m$, $n \geq 1, m \geq 1$. We shall assume that in (2.134), the functions $u(x, z)$ and

$$\lambda(x) = \int\limits_{R^m} F\big(x, z, u, u_x, u_z\big)dz, \quad \left(\frac{\partial\lambda}{\partial z_k} = 0, \ \ k = 1, 2, \ldots, m\right)$$

are unknown.

Assume the following conditions also hold

$$u\Big|_{z=0} = f(x) \quad \exists w(x,\omega) = \frac{1}{(2\pi)^m} \int\limits_{R^m} u(x,z) e^{iz\omega} dz. \tag{4.25}$$

Let $\mathcal{A}_x, \mathcal{B}_x$ be independent of z; L_z, D_z be operators with constant coefficients with symbols $l(\omega), d(\omega)$, $\omega \in R^m$ respectively.

Theorem 2.14. The following relations

$$\mathcal{A}_x w + l(\omega) w - \frac{\mathcal{A}_x f + \int\limits_{R^m} l(y) w(x,y) dy}{\mathcal{B}_x f + \int\limits_{R^m} d(y) w(x,y) dy} \Big[\mathcal{B}_x w + d(\omega) w\Big] = 0,$$

$$\lambda(x) = \frac{\mathcal{A}_x f + \int\limits_{R^m} l(y) w(x,y) dy}{\mathcal{B}_x f + \int\limits_{R^m} d(y) w(x,y) dy} \tag{4.26}$$

hold.

Now, in (2.134), let $x = (y,t)$, $D_z = 0$, $\mathcal{A}_x = -\frac{\partial}{\partial t} + \hat{\mathcal{A}}_y$, where $\hat{\mathcal{A}}_y$ is a linear operator independent of (z,t); $\mathcal{B}_y$ is an operator not depending on t.

Theorem 2.15. Let $\mathcal{B} f \neq 0$. Then

$$u(y,z,t) = \frac{f(y,t)\varphi(z,t)}{\varphi_0(t)} + R(y,z,t),$$

$$\lambda(y,t) = \frac{(f_t - \mathcal{A}_y f)\varphi_0 - f\varphi_0'(t)}{\varphi_0(t)\hat{\mathcal{B}}_y f},$$

where $\varphi(z,t)$, $R(y,z,t)$ are solutions to equations

$$\frac{\partial \varphi}{\partial t} = L_z \varphi, \quad \frac{\partial R}{\partial t} = \hat{\mathcal{A}}_y R + L_z R + \frac{(f_t - \mathcal{A}_y f)\varphi_0 - f\varphi_0'(t)}{\varphi_0(t) f} \mathcal{B} R,$$

such that

$$\varphi_0\Big|_{z=0} = \varphi_0(t) \neq 0, \qquad R\Big|_{z=0} = 0.$$

Numerous concrete results connected with boundary-value problems for integro-differential equations of type (2.132), (2.136), with the additional to (2.131), (2.135) information, may be found in Anikonov (1986, 1988), Anikonov and Belov (1989), Anikonov and Bubnov (1988a, b; 1989) and in the bibliography cited there.

Concrete choice of the functions $\varphi(z,t)$ and $R(y,z,t)$ in theorem 2.15 depends on information, additional to (2.135).

4. Below we shall give explicit formulas for solutions to inverse problems for certain nonlinear equations.

a) Let D_x^α be an operator of differentiation by x of the order m, $x \in R^n$, a_α be constants. Consider the nonlinear equation

$$\sum_{|\alpha|=m} a_\alpha D_x^\alpha \lambda\big(u(x,t)\big) = \frac{\partial^m u}{\partial t^m}, \quad m \geq 1.$$

Here $\lambda(u), u(x,t)$ are the desired functions. The formulas for $\lambda(u)$ and $u(x,t)$ will be given depending on an $m+1$ function of one variable.

Let $a(y)$, $\varphi^\alpha(y)$ be certain m times differentiable functions, $y \in R$. Set $\varphi^\alpha = (\varphi_1^{\alpha_1}, \varphi_2^{\alpha_2}, \dots, \varphi_n^{\alpha_n})$, $\alpha = (\alpha_1, \alpha_2, \dots, \alpha_n)$.

Theorem 2.16. The following formulas

$$\lambda^{-1}(y) = \int \sum_{|\alpha|=m} a_\alpha \varphi^\alpha\big(a^{-1}(y)\big)\,dy,$$

$$u(x,t) = \lambda\big(v(x,t)\big)$$

hold, where $v(x,t)$ is a root of the equation

$$v(x,t) = a\left(t + \sum_{j=1}^{n} x_j \varphi_j\Big(a^{-1}\big(v(x,t)\big)\Big)\right).$$

Ambiguities may arise here, as before, since the equation $a(y) = p$ has roots.

In conclusion, consider the inverse problem for the nonlinear second order equation: to find the function $\mu(y)$, $y \geq 0$ if

$$\frac{d^2 u}{dt^2} = R\mu(u), \quad 0 \leq t \leq 1,$$

$$u(0,k) = u(1,k) = 0, \qquad \left.\frac{\partial u}{\partial t}\right|_{t=0} = a(k).$$

Theorem 2.17. The formula

$$\mu(y) = \frac{1}{2}\frac{d}{dy}\lambda^2(y),$$

holds, where $\lambda(y)$ is inverse to

$$y(p) = \frac{1}{\pi}\int\limits_p^1 \frac{q æ(p)dq}{\sqrt{q^2 - p^2}},$$

and $æ(p)$ is inverse to

$$f(x) = 1 - \frac{a^2(x^2)}{x^2}.$$

Remark. As follows from above, the essential moment for inverse problem solving is inverting the mapping connected with initial information distributed in the given space. Therefore, it seems that such sections of mathematics as topology, complex analysis, catastrophe theory, and others may be applied to solve these problems.

§2.5 Inverse problems of transfer theory.

Initial boundary-value problems for transfer equations were investigated by many authors. A quite extensive bibliography may be found in Aleksandrov (1937) and Anikonov (1968, 1969, 1978).

In this paragraph we shall set the inverse problems for the transfer equation. The results will relate to both the uniqueness and existence of solutions to these problems.

Consider the boundary-value problem for the transfer equation

$$u_t + (\omega, \nabla_x u) + \sigma(x,\omega)u$$

$$= \sigma_s(x,\omega) \int_\Omega g(x,\omega,\omega')u(t,x,\omega')d\omega' + h(t,x,\omega)f(x,\omega), \tag{5.1}$$

$$u\Big|_{\Gamma_-} = \varphi, \tag{5.2}$$

where $t \in R$, $x \in \mathcal{D} \subset R^n$, $\mathcal{D}$ is a strictly convex bounded domain with a boundary $\gamma \in C^2$; $\omega \in \Omega$, Ω is a unit sphere in R^n;

$$\Gamma_- = \{(x,t,\omega') : t \in R, \quad x \in \gamma, \quad (n,\omega) < 0\};$$

n is an external unit vector, normal to γ.

We shall set three inverse problems.

1) To find functions $u(x,t,\omega)$, $\sigma(x,\omega)$.

2) To find functions $u(x,t,\omega)$, $\sigma_s(x,\omega)$.

3) To find functions $u(x,t,\omega)$, $f(x,\omega)$ entering into (2.137) if parallel with (2.137), (2.138) we add the data for $t = 0$ and the conditions imposed on $u(t,\omega,x)$ with respect to t

$$u\Big|_{t=0} = u_0, \quad u_0\Big|_{\gamma_-} = \varphi\Big|_{t=0}, \quad \gamma_- = \Big\{(x,\omega) : x \in \gamma, \ (n,\omega) < 0\Big\}; \tag{5.3}$$

$$\max_{x,\omega} \int_{-\infty}^{+\infty} \left(1 + |\lambda|^{2p}\right) \left[|\hat{u}|^2 + |\nabla_x \hat{u}|^2\right] d\lambda < \infty, \quad p \geq 2, \tag{5.4}$$

Here, $\hat{u}$ is the Fourier transform of $u(t,x,\omega)$ with respect to t. The well-posedness of these settings is guaranteed by the following theorem.

Theorem 2.18. Let

$$\sigma, \sigma_s \in C^1(\bar{\mathcal{D}}) \times C(\Omega), \quad g \in C^1(\bar{\mathcal{D}}) \times C(\Omega) \times C(\Omega)$$

and the following condition

$$\sigma d^2 \left[\left(\left\|\sigma\right\|_0 + \left\|\sigma_s\right\|_0 \cdot \left\| \int_\Omega |g| d\omega' \right\|_0^2\right)\right.$$

$$+\left(\left\|\nabla_x\sigma\right\|_0+\left\|\nabla_x\sigma_s\right\|_0\left\|\int\limits_\Omega\left|g\right|d\omega'\right\|_0^2+\left\|\sigma_s\right\|_0\left\|\int\limits_\Omega\left|\nabla_x g\right|d\omega'\right\|_0^2\right)\Bigg]\leq 1$$

hold. Then for any functions hf and φ such that

$$E\left(\hat{\varphi}_0,\hat{h}f\right)=\max_{\omega,x\in\gamma}\int\limits_{-\infty}^{+\infty}\left(1+|\lambda|^{2p}\right)\left[|\hat{\varphi}|^2+|\nabla_x\hat{\varphi}|^2\right]d\lambda$$

$$+\max_{\omega,x}\int\limits_{-\infty}^{+\infty}\left(1+|\lambda|^{2p}\right)\left[|\hat{f}h|^2+|\nabla_x\hat{f}h|^2\right]d\lambda<\infty,$$

there exists a unique solution to problem (2.137), (2.138) and the estimate

$$\max_{\omega,x\in\gamma}\int\limits_{-\infty}^{+\infty}\left(1+|\lambda|^{2p}\right)\left[|\hat{u}|^2+|\nabla_x\hat{u}|^2\right]d\lambda\leq cE(\hat{\varphi},\hat{h}f),\quad p\geq 2,$$

$$\left\|\varphi(x,\omega)\right\|_0=\max_{x,\omega}\left|\varphi(x,\omega)\right|,\quad d=\operatorname{diam}\mathcal{D}$$

holds.

The inverse problems 1), 2), 3) may be reduced to investigating boundary-value problems for integro-differential equations. This reduction is formulated in the lemmas below.

Lemma 2.5. If $v(\lambda,x,\omega)$ solves the boundary-value problem

$$-i\lambda v+\left(\omega,\nabla_x v\right)+v\left[F_0(x,\omega)+\frac{i}{u_0}\int\limits_{-\infty}^{+\infty}\lambda v\,d\lambda\right]=\sigma_s(x,\omega)\int\limits_\Omega gv\,d\omega+f\hat{h},\tag{5.5}$$

$$v\Big|_{\gamma_-}=\hat{\varphi},\quad |u_0|\geq\sigma>0,\quad \lambda\in R\tag{5.6}$$

from the class of functions v such that

$$\max_{\omega,x}\int\limits_{-\infty}^{+\infty}\left(1+|\lambda|^{2p}\right)\left[|v|^2+|\nabla_x v|^2\right]d\lambda<\infty,\quad p\geq 2,$$

then the desired functions $u(t,x,\omega)$, $\sigma(x,\omega)$ in inverse problem 1) are calculated by the formulas

$$\sigma(x,\omega)=\frac{1}{u_0}F_0(x,\omega)+\frac{1}{u_0}\int\limits_{-\infty}^{+\infty}\lambda v d\lambda,\quad u(t,x,\omega)=\int\limits_{-\infty}^{+\infty}ve^{-i\lambda t}d\lambda,$$

$$F_0(x,\omega)=\sigma_s(x,\omega)\int\limits_{\Omega}gu_0d\omega'-h(0,x,\omega)f(x,\omega)-\left(\omega,\nabla_x u_0\right).$$

Lemma 2.6. If $v(\lambda,x,\omega)$ solves the boundary-value problem

$$-i\lambda v+\left(\omega,\nabla_x u_0\right)+\sigma v=F_1-\frac{i}{\int\limits_{\Omega}gu_0d\omega}\cdot\int\limits_{-\infty}^{+\infty}\lambda v d\lambda\int\limits_{\Omega}gvd\omega'+f\hat{h},\quad (5.7)$$

$$v\Big|_{\gamma_-}=\hat{\varphi},\qquad\left|\int\limits_{\Omega}gu_0d\omega'\right|\geq\delta>0\qquad(5.8)$$

from the class of functions v such that

$$\max_{\omega,x}\int\limits_{-\infty}^{+\infty}\left(1+|\lambda|^{2p}\right)\left[|v|^2+|\nabla_x v|^2\right]d\lambda<\infty,\quad p\geq 2,$$

then the desired functions $u(t,x,\omega)$, $\sigma_s(x,\omega)$ from inverse problem 2) are calculated by the formulas

$$\sigma_s(x,\omega)=F_1-\frac{i}{\int\limits_{\Omega}gu_0d\omega}\cdot\int\limits_{-\infty}^{+\infty}\lambda v d\lambda,$$

$$u(t,x,\omega)=\int\limits_{-\infty}^{+\infty}v(\lambda,x,\omega)\exp(-i\lambda t)d\lambda,$$

$$F_1(x,\omega)=\left[\left(\omega,\nabla_x u\right)+\sigma u_0-fh(0,x,\omega)\right]\left|\int\limits_{\Omega}gu_0d\omega'\right|.$$

Lemma 2.7. If $v(\lambda, x, \omega)$ solves the boundary-value problem

$$-i\lambda v + \Big(\omega, \nabla_x v\Big) + \sigma v = \sigma_s \int\limits_\Omega g v d\omega'$$

$$+ \frac{\hat{h}(\lambda, x, \omega)}{h(0, x, \omega)} \left[-i \int\limits_{-\infty}^{+\infty} \lambda v d\lambda + \Big(\omega, \nabla_x u_0\Big) + \sigma u_0 - \sigma_s \int\limits_\Omega g u_0 d\Omega' \right], \quad (5.9)$$

$$v\Big|_{\gamma_-} = \hat{\varphi}, \quad \Big|h(0, x, \omega)\Big| \geq \delta > 0 \quad (5.10)$$

from the class of functions v such that

$$\max_{\omega, x} \int\limits_{-\infty}^{+\infty} \Big(1 + |\lambda|^{2p}\Big) \Big[|v|^2 + |\nabla_x v|^2\Big] d\lambda < \infty, \quad p \geq 2,$$

then the desired functions $u(t, x, \omega)$, $f(x, \omega)$ from inverse problem 3) may be calculated by the formulas

$$f(x, \omega) = \frac{1}{h(0, x, \omega)} \left[\Big(\omega, \nabla_x u_0\Big) + \sigma u_0 - \sigma_s \int\limits_\Omega g u_0 d\omega' - i \int\limits_{-\infty}^{+\infty} \lambda v d\lambda \right],$$

$$u(t, x, \omega) = \int\limits_{-\infty}^{+\infty} v(\lambda, x, \omega) \exp(-i\lambda t) dt.$$

Solvability of problems (2.141), (2.142)-(2.145), (2.146) is established in the following theorems.

Theorem 2.19. Let $\sigma_s \in C^1(\bar{\mathcal{D}}) \times C(\Omega)$, $g \in C^1(\bar{\mathcal{D}}) \times C(\Omega) \times C(\Omega)$, $u_0 \in C^2(\bar{\mathcal{D}}) \times C(\Omega)$, $|u_0| \geq \sigma$, $h(0, x, \omega) f(x, \omega) \in C^1(\bar{\mathcal{D}}) \times C(\Omega)$, $E(\hat{\varphi}, \hat{h} f) < \infty$, $p \geq 2$ and the condition

$$d^2 \left(\frac{\Big\|F_0\Big\|}{\delta^2} + \Big\|\sigma_s\Big\|_0^2 \Big\| \int\limits_\Omega |g| d\omega' \Big\|_0^2 \right) + \frac{(12)^3 d^2 R_1^2}{\delta^2 (2p - 3)}$$

$$\frac{4(12)^2 d^2 R_1^2}{\delta^2 (2p-3)} \Big\|\nabla u_0\Big\|_0^2 + 12 d^2 \left[\frac{\Big\|F_0\Big\|_0 \Big\|\nabla_x u_0\Big\|_0}{\delta^2} + \frac{\Big\|\nabla_x F_0\Big\|}{\delta} \right.$$

$$+\left\|\nabla_x\sigma_s\right\|\left\|\int_\Omega |g|d\omega'\right\|_0+\left\|\sigma_s\right\|_0\left\|\int_\Omega |\nabla_x y|d\omega'\right\|_0\Bigg]^2 \leq 1,$$

$$R_1^2=\max_{x\in\gamma_-,\omega}\int_{-\infty}^{+\infty}(1+|\lambda|)^{2p}\left[|\hat\varphi|^2+|\nabla_x\hat\varphi|^2\right]d\lambda$$

$$+d^2\max_{x,\omega}\int_{-\infty}^{+\infty}(1+|\lambda|)^{2p}\left[|\nabla_x\hat h f|^2+|\hat h f|^2\right]d\lambda,\quad p\geq 2$$

hold. Then there exists a unique solution to problem (2.141), (2.142) and the estimate

$$\max_{x,\omega}\int_{-\infty}^{+\infty}(1+|\lambda|)^{2p}\left[|v|^2+|\nabla_x v|^2\right]d\lambda\leq cF(\hat\varphi,\hat h f)$$

holds.

Theorem 2.20. Let $\sigma\in C^1(\bar{\mathcal D})\times C(\Omega)$, $g\in C^1(\bar{\mathcal D})\times C(\Omega)\times C(\Omega)$, $u_0\in C^2(\bar{\mathcal D})\times C(\Omega)$, $\left|\int_\Omega g u_0 d\omega'\right|\geq\delta>0$, $fh(0,x,\omega)\in C^1(\bar{\mathcal D})\times C(\Omega)$, $E(\hat\varphi,\hat h f)<\infty$, φ, hf be real functions and the condition

$$12d^2\left(\left\|\sigma\right\|_c+\left\|F_1\right\|_0\left\|\int_\Omega |g|d\omega'\right\|_0^2\right)$$

$$+\frac{(12)^3R_1^2d^2}{\delta^2(2p-3)}\left\|\int_\Omega |g|d\omega'\right\|_0^2$$

$$+12d^2\Bigg(\left\|\nabla_x\sigma\right\|_0+\left\|\nabla_x F_1\right\|_0\cdot\left\|\int_\Omega |g|d\omega'\right\|_0$$

$$+\left\|F_1\right\|_0\left\|\int_\Omega |\nabla_x g|d\omega'\right\|_0\Bigg)^2$$

$$+\frac{2(12)^2R_1^2d^2}{\delta^4(2p-3)}\left\|\int_\Omega |g|d\omega'\right\|_0^2\left\|\int_\Omega |\nabla_x g u_0|d\omega\right\|_0^2$$

$$+\frac{(12)^3R_1^2d^2}{\delta^2(2p-3)}\Big\|\int\limits_\Omega|\nabla_x g|d\omega\Big\|_0^2\leq 1,\quad p\geq 2$$

hold. Then there exists a unique solution to problem (2.143), (2.144) and the estimate

$$\max_{x,\omega}\int\limits_{-\infty}^{+\infty}(1+|\lambda|)^{2p}\left[|v|^2+|\nabla_x v|^2\right]d\lambda\leq cE(\hat{\varphi},\hat{h}f),\quad p\geq 2$$

holds.

Theorem 2.21. Let $\sigma,\sigma_s\in C^1(\bar{\mathcal{D}})\times C(\Omega)$, $g\in C^1(\bar{\mathcal{D}})\times C(\Omega)\times C(\Omega)$, $u_0\in C^2(\bar{\mathcal{D}})\times C(\Omega)$, $h(0,x,\omega)\neq 0$, $E(\hat{\varphi},\hat{h})<\infty$, $p\geq 2$ and the condition

$$6d^2\left\{\left[\|\sigma\|_0+\|\sigma_s\|_0\Big\|\int\limits_\Omega|g|d\omega'\Big\|_0\right.\right.$$
$$\left.+\frac{1}{|h(0,x,\omega)|}\left(\int\limits_{-\infty}^{+\infty}\frac{(1+|\lambda|^{2p})\,|h|d\lambda)}{2p-3}\right)^{\frac{1}{2}}\right]$$
$$+\left(\|\nabla_x\sigma\|_0+\|\nabla_x\sigma_s\|_0\Big\|\int\limits_\Omega|g|d\omega'\Big\|_0\right.$$
$$\left.\left.+\|\sigma_s\|_0\Big\|\int\limits_\Omega|\nabla_x g|d\omega'\Big\|_0\right)^2\right\}\leq 1$$

hold. Then there exists a unique solution to problem (2.145), (2.146) and the estimate

$$\max_{x,\omega}\int\limits_{-\infty}^{+\infty}(1+|\lambda|)^{2p}\left[|v|^2+|\nabla_x v|^2\right]d\lambda\leq cE(\hat{\varphi},\hat{h}(\lambda)),\quad p\geq 2$$

holds.

The proof of these theorems is based on investigation of nonlinear integral equations equivalent to the corresponding boundary-value problems. Note, that the conditions of theorems 2.18-2.21 hold when the measure of $\mathcal{D}$ is sufficiently small, or when the norms of the given data are sufficiently small.

§2.6 Geometric problems.

Further we shall look for the convex domain $B \in R^3$. The given information will be the function

$$f(n) = \int_B K(n,p)b(p)dS_p,$$

where

$$K(n,p) = a|\langle n,p\rangle| + b\chi(n,p),$$

a, b are constants, $\langle n,p\rangle$ is the scalar product; χ is the Heaviside function; $b(p)$ is a certain function; $|n| = 1$; $|p| = 1$.

In this paragraph, we recover the manifolds (the sources) by the functionals of their orthogonal projections and their illuminated parts.

The integral equation of the first kind. We consider here the integral equation of the first kind. It arises when we try to determine the convex domain by the functionals of its orthogonal projections (shadows) and illuminated parts. Some of the results below are contained in Anikonov (1978), Anikonov and Golubyatnikov (1978), Anikonov and Marchuk (1975), Anikonov and Stepanov (1981).

Let x be a point of the unit sphere $\omega = \{x \in R^3 : |x| = 1\}$, $K(n,p) = a|\langle n,p\rangle| + b\chi(\langle n,p\rangle)$; a, b be constants; χ be the Heaviside function; $\langle n,p\rangle$ be the scalar product; $\omega(n) = \{p \in \omega;\ \langle n,p\rangle \geq 0\}$; $n \in \omega$, the hemisphere depending on p.

Consider the first order integral equation relative to the continuous function $z(p)$, $p \in \omega$

$$f(n) = \int_\omega K(n,p)z(p)d\omega_p = a\int_\omega |\langle n,p\rangle|\, z(p)d\omega_p + b\int_{\omega(n)} z(p)d\omega_p. \quad (6.1)$$

The uniqueness of the solution to this equation is established by theorems 2.22-2.23.

Theorem 2.22. If $a \neq 0$, $b \neq 0$, $a + b \neq 0$, then (2.147) has no more than one continuous solution $z(p)$, $p \in \omega$.

Theorem 2.23. If the function $f(n)$ from (2.147) is three times continuously differentiable and the conditions of theorem 2.1 hold, then the inversion formula

$$z(n) = -\frac{1}{8\pi^2}\frac{d}{dt}\int\limits_{\langle n,p\rangle^2 > t} \frac{(\Delta + 2)f_1(p)\,|\langle n,p\rangle|}{\sqrt{\langle n,p\rangle^2 - t}}\,d\omega_p\Big|_{t=0} - \frac{1}{4\pi^2 b}\int\limits_{\omega(n)} \frac{\Delta f^-(p)}{\langle n,p\rangle}\,d\omega_p, \tag{6.2}$$

holds, where

$$f_1(p) = \frac{1}{a}f^+(p) - \frac{b}{4\pi a(a+b)}\int\limits_{\omega} f(p)\,d\omega_p;$$

$$f^+(p) = \frac{f(p) + f(-p)}{2}; \quad f^-(p) = \frac{f(p) - f(-p)}{2};$$

Δ is the Laplace-Beltrami operator on ω. For the particular case when, in integral equation (2.147), the kernel is $K(n,p) = |\langle n,p\rangle|$ and $z(p)$ is even with respect to ω, the inversion formula is given in Blyashke (1967) and Pogorelov (1973).

Theorem 2.24. The eigenvalues λ of equation (2.147) are as follows

$$\lambda_r = \begin{cases} 2\pi(a+b), & r = 0 \\ 4\pi a(-1)^{k+1}\frac{(12k-3)!!}{(2k+2)!!}, & r = 2k, \quad k = 0,1,2,\ldots, \\ 2b(-1)^k\frac{(12k-1)!!}{(2k+2)!!}, & r = 2k+1, \quad k = 0,1,2,\ldots, \end{cases}$$

and the eigen subspace corresponding to λ_r consists of all spherical harmonics of the order r.

Regularity of solution $z(p)$ and its stability is established by theorems 2.25, 2.26. So, using the fact of the exponential decrease of Fourier decomposition terms of the analytic in ω function (Sobolev, 1974), we obtain theorem 2.25.

Theorem 2.25. Let conditions of theorem 2.22 hold. Then the solution $z(n)$ to equation (2.147) is analytic in ω if and only if $f(n)$ is analytic in ω.

Analysis of inversion formula (2.148) leads to the following result.

Theorem 2.26. If $a \neq 0$, $b \neq 0$, $a + b \neq 0$ and $f(n) \in C^k(\omega)$, $k \geq 4$, then the solution $z(n)$ to equation (2.147) belongs to the class $C^{k-4}(\omega)$ and the estimate

$$\left\|z(n)\right\|_{C(\omega)} \leq A\left(\frac{1}{|a|} + \frac{1}{|b|} + \frac{|b|}{|a(a+b)|}\right)\left\|f(n)\right\|_{C^4(\omega)}, \tag{6.3}$$

where A is a constant independent of f.

Let μ be the Lebesque measure in the sphere ω. Consider the first order equation relative to this measure

$$f(n) = a\int\limits_{\omega} |\langle n, p\rangle|\, \mu(d\omega_p) + b\int\limits_{\omega(n)} \mu(d\omega_p), \quad n \in \omega. \tag{6.4}$$

Theorem 2.27. If $a \neq 0$, $b \neq 0$, $a + b \neq 0$, then equation (2.150) has no more than one solution.

On the basis of the inverse formula (theorem 2.23) we may in the standard way investigate more general integral equations of the first order, among them, nonlinear equations. Suppose, for example, we consider the equation

$$f(n) = \int\limits_{\omega} K(n,p)z(p)d\omega_p + (Az)(n) \equiv Kz + Az,$$

where

$$K(n,p) = a\,|\langle n,p\rangle| + b\chi\,|\langle n,p\rangle|\,,$$

and A is a certain operator. By the inversion formula we have

$$z(p) = \hat{f}(n) + (Bz)(n), \tag{6.5}$$

where $\hat{f} = K^{-1}f$, $B = -K^{-1}A$. Relation (2.151) is a second order equation. For example, for corresponding restrictions imposed on A, the operator B may be contractive, which allows us to use the method of fixed point theory.

Another application of the obtained formula is connected with problems of integral geometry type. Let $\Gamma(n)$ be the circumference of a great circle of the sphere ω, orthogonal to the vector $n \in \omega$; Δ be the second Beltrami differential operator in the sphere ω.

Consider the following problem. To find a two times continuously differentiable function $u(p)$, $p \in \omega$ such that

$$g(n) = 2a \int\limits_{\Gamma(n)} u(p)ds + b \int\limits_{\omega(n)} (\Delta + 2)u(p)d\omega_p.$$

Here a, b are constants, $a \neq 0$, $b \neq 0$, $a+b \neq 0$; ds is the length element of $\Gamma(n)$.

Theorem 2.28. Let $g(n)$ be four times continuously differentiable. Then the following formula holds

$$(\Delta + 2)u(n) = -\frac{1}{8\pi^2}\frac{d}{dt} \int\limits_{\langle n,p\rangle^2 > t} \frac{(\Delta+2)g(p)\,|\langle n,p\rangle|}{\sqrt{\langle n,p\rangle^2 - t}}d\omega_p \Bigg|_{t=0}$$

$$- \frac{1}{4\pi^2 b} \int\limits_{\omega(n)} \frac{\Delta g^-(n)}{\langle n,p\rangle}d\omega_p,$$

$$g_1(p) = \frac{1}{a}g^+(p) - \frac{b}{4\pi(a+b)} \int\limits_{\omega} g(p)d\omega_p,$$

$$g^+(p) = \frac{g(p)+g(-p)}{2}, \quad g^-(p) = \frac{g(p)-g(-p)}{2}.$$

Uniqueness. We shall give here some theorems on unique definiteness of a closed convex surface, if the functionals of its orthogonal projections and illuminated parts are known. The method is as follows: first, the problem is reduced to integral equation (2.147); then we use the results in the Minkovskii and Kristoffel problems and their generalizations. All surfaces are assumed, as a rule, to be strictly convex and sufficiently smooth. Some of results represented here may be found in (Alekseev, 1962; Anikonov, 1968, 1969; Anikonov and Stepanov, 1981; Stepanov, 1981, 1984).

Consider the function $\varphi(n) = 2aF(n) + bS(n)$, $n \in \omega$, where $F(n)$ is the area of orthogonal projection of a closed convex surface B in the plane

$$P(n) = \left\{x \in R^3;\ \langle x, n\rangle = 0\right\};$$

$S(n)$ is the illuminated area in the direction n to the surface B; a, b are constants, $a \neq 0$, $b \neq 0$, $a + b \neq 0$.

For $F(n)$ and $S(n)$, the following representations hold true

$$F(n) = \frac{1}{2}\int\limits_{\omega} |\langle n, p\rangle|\, G(B, d\omega_p), \quad S(n) = \int\limits_{\omega(n)} G(B, d\omega_p),$$

where $G(B, \cdot)$ is the surface function of B. Using these representations we may write out the function $\varphi(n)$

$$\varphi(n) = a\int\limits_{\omega} |\langle n, p\rangle|\, G(B, d\omega_p) + b\int\limits_{\omega(n)} G(B, d\omega_p). \tag{6.6}$$

A closed convex surface is uniquely determined to within the parallel transfer by the surface function (Aleksandrov, 1937). Therefore, since solution to (2.152) is unique, theorem 2.27 holds.

Theorem 2.29. A closed convex surface is uniquely determined to within the parallel transfer by the function $\varphi(n)$, $n \in \omega$.

Let $G(B, \cdot)$ be the first curvature function of a closed convex surface B; $L(n)$ be the length of the boundary of orthogonal projection of the surface B in the plane $P(n)$; $M(n)$ be the mean integral curvature of the illuminated surface of B in the direction n. The functions $L(n)$, $M(n)$ are determined as follows

$$L(n) = \frac{1}{2}\int_{\omega} |\langle n,p\rangle|\, G_1(B, d\omega_p), \quad M(n) = \frac{1}{2}\int_{\omega(n)} G_1(B, d\omega_p).$$

Consider the function

$$\psi(n) = 2aL(n) + 2bM(n),$$

where a, b are constants, $a \neq 0$, $b \neq 0$, $a + b \neq 0$. Taking into account the representations for $L(n)$ and $M(n)$ we may write

$$\psi(n) = a\int_{\omega} |\langle n,p\rangle|\, G_1(B, d\omega_p) + b\int_{\omega(n)} G_1(B, d\omega_p). \tag{6.7}$$

This is a first order equation relative to the curvature function $G_1(B,\cdot)$. By theorem 2.27, the solution to this equation is unique. As a closed convex surface is uniquely determined to within a parallel transfer by the first curvature function, then the following theorem holds.

Theorem 2.30. A closed convex surface B is uniquely determined to within a parallel transfer by the function $\psi(n)$, $n \in \omega$.

Remark. If a surface B is centrally symmetric, then theorems 2.29 and 2.30 hold also for $b = 0$. This means that each convex centrally symmetric surface B is uniquely determined by areas and lengths of boundaries of its orthogonal projections in the planes in all directions. This was known before (Aleksandrov, 1937; Blyashke, 1967). If a surface is not centrally symmetric, this statement does not hold. So, the piecewise analytic surface of rotation, the generator of which has the parametric representation

$$x = \frac{A}{\sqrt{2}}\cos\varphi, \quad y = 0, \quad x = A\cos\left(\frac{\pi}{4} - \varphi\right), \quad y = 0;$$

$$z = A\int\limits_0^{\varphi}\sqrt{1-\frac{\sin^2\varphi}{2}}d\varphi;\ 0\le\varphi\le\frac{\pi}{2},\ \ z = A\left(\frac{1}{\sqrt{2}}-\sin\left(\frac{\pi}{2}-\varphi\right)\right),$$

$$-\frac{\pi}{4}\le\varphi\le 0,$$

is, like the sphere, a surface of constant brightness, i.e. it has a constant projection area in each plane (Blyashke, 1967). The lengths of boundaries of all orthogonal projections also do not determine a convex surface that is not centrally symmetric. For example, the convex surface with the support function

$$H(x) = |x| + h(x),$$

where $h(x)$ is even, is a surface of constant scope, i.e., it has a constant length of projection in each plane.

Theorems 2.29, 2.30 do not provide the most effective method of determining the form of a surface. In the case when B is sufficiently smooth, the problem of determining its form by $\varphi(n)$ or $\psi(n)$ is reduced to inversion of (2.152), (2.153) and to solving the second order equation with partial derivatives for the surface support function.

Let $R_1(n)$, $R_2(n)$ be the principal radii of curvature of a closed, strictly convex, two times continuously differentiable surface B, as the functions of the unit normal vector n. In this case the surface function $G(B,\cdot)$ has the continuous density $z(n) = R_1(n)R_2(n)$ and the function $\varphi(n)$ is as follows

$$\varphi(n) = a\int\limits_{\omega}|\langle n,p\rangle|\,R_1(p)R_2(p)d\omega_p + b\int\limits_{\omega(n)}R_1(p)R_2(p)d\omega_p. \qquad (6.8)$$

Consider (2.154) as the first order equation relative to the product $R_1(n)R_2(n)$.

Theorem 2.31. Let $a\neq 0$, $b\neq 0$, $a+b\neq 0$; B be a closed strictly convex regular surface. Then

$$R_1(n)R_2(n) = -\frac{1}{8\pi^2}\frac{d}{dt}\int\limits_{\langle n,p\rangle^2>t}\frac{(\Delta+2)\varphi_1(p)\,|\langle n,p\rangle|}{\sqrt{\langle n,p\rangle^2-t}}d\omega_p\Bigg|_{t=0} - \frac{1}{4\pi^2 b}\int\limits_{\omega(n)}\frac{\Delta\varphi^-(p)}{\langle n,p\rangle}d\omega_p,$$

$$R_1(n)\cdot R_2(n) = \begin{vmatrix} H_{11}(n) & H_{12}(n) \\ H_{21}(n) & H_{22}(n) \end{vmatrix} + \begin{vmatrix} H_{22}(n) & H_{23}(n) \\ H_{32}(n) & H_{33}(n) \end{vmatrix} + \begin{vmatrix} H_{33}(n) & H_{13}(n) \\ H_{31}(n) & H_{11}(n) \end{vmatrix}, \qquad (6.9)$$

where $H_{ij}(n)$ are the derivatives $\frac{\partial^2 H(x)}{\partial x_i \partial x_j}$ of the support function $H(x)$ taken for $x = n \in \omega$.

It follows from this theorem that reconstruction of a regular surface B by $\varphi(n)$ is reduced to solving (2.154) for its support function. Note that (2.155) is already well-known (Bakel'man *et al.*, 1973).

For the above conditions imposed on B, the first curvature function $G_1(B)$ has a density equal to the sum of the principal radii of curvature; therefore, $\psi(n)$ may be represented as follows

$$\psi(n) = a\int\limits_{\omega}|\langle n,p\rangle|\,[R_1(p)+R_2(p)]\,d\omega_p + b\int\limits_{\omega(n)}[R_1(p)+R_2(p)]\,d\omega_p.$$

Theorem 2.32. Let $a \neq 0$, $b \neq 0$, $a+b \neq 0$, B be a closed strictly convex regular surface. Then

$$R_1(n)+R_2(n) = -\frac{1}{8\pi^2}\frac{d}{dt}\int\limits_{\langle n,p\rangle^2>t}\frac{(\Delta+2)\psi_1(p)\,|\langle n,p\rangle|}{\sqrt{\langle n,p\rangle^2-t}}d\omega_p\Bigg|_{t=0} - \frac{1}{4\pi^2 b}\int\limits_{\omega(n)}\frac{\Delta\psi^-(p)}{\langle n,p\rangle}d\omega_p,$$

$$H(n) = \frac{1}{4\pi}\int\limits_{\omega}[R_1(p)+R_2(p)]\,(1-\langle n,p\rangle)\ln(1-\langle n,p\rangle)\,d\omega_p. \qquad (6.10)$$

Using (2.156), we may reconstruct a regular surface by $\psi(n)$ explicitly. This formula (2.156) is contained in Pogorelov (1969).

Now consider the problems of determining a closed convex surface and a function given in this surface if we know the functionals of the orthogonal projections and of illuminated parts.

Let B be a closed strictly convex smooth surface in R^3, and $u(n)$ be a continuous function in B, as the function of unit normal vector n. Consider the functions

$$\psi_1(n) = a_1 \int\limits_{\omega} |\langle n, p\rangle|\, u(p) R_1(p) R_2(p) d\omega_p + b_1 \int\limits_{\omega(n)} u(p) R_1(p) R_2(p) d\omega_p,$$

$$\psi_2(n) = a_2 \int\limits_{\omega} |\langle n, p\rangle|\, u(p) \left[R_1(p) + R_2(p)\right] d\omega_p$$

$$+ b_2 \int\limits_{\omega(n)} u(p) \left[R_1(p) + R_2(p)\right] d\omega_p,$$

where $R_i(p)$ are the principal radii of curvature of B; a_i, b_i are constants; $a_i \neq 0$, $b_i \neq 0$, $a_i + b_i \neq 0$, $i = 1, 2$. The geometric sense of these functions for $u(p) = 1$ was given before.

Theorem 2.33. A closed strictly convex, two times continuously differentiable surface B and a continuous function $u(n) > 0$, $n \in \omega$ are uniquely determined by the functions $\varphi_1(n)$, $\varphi_2(n)$, $n \in \omega$ (i.e., B is reconstructed to within a parallel transfer).

In dispersion problems or in radiation problems we have to determine both the dispersing object $B(t)$, the form of which may change, and the function of the source density in it. Bearing this in mind, we consider the equation

$$\Phi(n, t) = \int\limits_{B(t)} K(n, p) u(p) dS_t,$$

where

$$K(n, p) = a\, |\langle n, p\rangle| + b\chi\left(\langle n, p\rangle\right);$$

dS_t is the element of the area $B(t)$.

Theorem 2.34. For each $t \geq 0$, let the surface be closed, strictly convex and two times continuously differentiable. Furthermore, let

$$\frac{d}{dt}\left[R_1(p,t)R_2(p,t)\right]\Big|_{t=0} = \alpha\left[R_1(p,0) + R_2(p,0)\right],$$

where $R_1(p,t)$, $R_2(p,t)$ are the principal radii of curvature of the surface $B(t)$; $\alpha \neq 0$ is a constant. Then the continuous function $u(p) > 0$ and the form of $B(t)$ are uniquely determined by the function $\Phi(n,t)$, $n \in \omega$, $t \geq 0$.

Suppose an area is given, together with a length of the boundary of orthogonal projection of a convex surface B in the planes of all possible directions. The following theorem holds.

Theorem 2.35. Let B_1 and B_2 be closed convex analytic surfaces in R^3. If $F_1(n) = F_2(n)$, $L_1(n) = L_2(n)$ for each $n \in \omega$, then B_1 is equal to B_2.

The condition of analyticity in this theorem is essential (Stefano Campi, 1986).

If B is strictly convex and two times continuously differentiable then the following theorem holds (Boonnesen and Fenchel, 1978).

Theorem 2.36. If $F(n) = const$, $L(n) = const$, then B is a sphere.

Existence. Further, we shall formulate the existence theorems for a closed convex surface with the given functionals of its orthogonal projections and illuminated parts.

Theorem 2.37. Let $F(n)$ be a continuous, strictly positive, even function in the unit sphere ω. Then there exists a closed, centrally symmetric surface B, unique to within a parallel transfer, for which the function $F(n)$ is the area of its orthogonal projection in the plane $\langle x, n\rangle = 0$. If $F(n)$ is k times continuously differentiable, $k \geq 4$ (analytic), then B is continuously differentiable at least $(k-3)$ times (analytic).

The proof of existence of B follows from the following two results.

1. There exists a unique symmetric measure μ in ω such that $F(n)$ is as follows

$$F(n) = \frac{1}{2\pi} \int_{\omega} |\langle n, p \rangle| \, \mu(d\omega_p).$$

2. This measure μ (as follows from Aleksandrov, 1937) is realized in a closed convex surface as its surface function.

To prove the regularity of B, we use the inversion formula for the integral equation

$$F(n) = \frac{1}{2} \int_{\omega} |\langle n, p \rangle| \, R_1(p) R_2(p) d\omega_p$$

and the results from regular solution of the Minkovskii problem (Pogorelov, 1975).

Theorem 2.38. In a sphere ω, a regular (k times continuously differentiable, $k \geq 7$, or analytic) strictly positive, even function $L(n)$ is given such that

$$I(n) \leq T''_{ss} \leq 0; \quad n \in \omega,$$

where

$$I(n) = \frac{d}{dt} \int_{\langle n,p \rangle^2 > t} \frac{(\Delta + 2) L(p) \, |\langle n, p \rangle|}{\sqrt{\langle n, p \rangle^2 - t}} d\omega_p \Bigg|_{t=0};$$

the index s denotes differentiation in the point $n \in \omega$ by the arcs of any great circle. Then there exists a unique, to within the parallel transfer, regular (of the class $C^{k-3,\alpha}$, $0 < \alpha < 1$, or, respectively, analytic) closed convex, centrally symmetric surface B; the length of orthogonal projection of B in the plane $\langle x, n \rangle = 0$ is $L(n)$.

Theorem 2.39. In the unit sphere ω in R^3, a regular (k times continuously differentiable, $k \geq 4$, or analytic) function $\varphi(n)$ is given such that

1) $n\varphi(n) d\omega_n = 0$,

2)

$$-\frac{1}{8\pi^2}\frac{d}{dt}\int\limits_{\langle n,p\rangle^2>t}\frac{(\Delta+2)\varphi_1(p)\,|\langle n,p\rangle|}{\sqrt{\langle n,p\rangle^2-t}}d\omega_p\Bigg|_{t=0}-\frac{1}{4\pi^2 b}\int\limits_{\omega(n)}\frac{\Delta\varphi^-(p)}{\langle n,p\rangle}d\omega_p>0,$$

where

$$\varphi_1(p)=\frac{1}{a}\varphi^+(p)-\frac{b}{4\pi a(a+b)}\int\limits_{\omega}\varphi(p)d\omega_p;$$

a, b are constants, $a\neq 0$, $b\neq 0$, $a+b\neq 0$. Then there exists a regular (at least $k-3$ times continuously differentiable, or, respectively, analytic) closed convex surface B for which the function $\varphi(n)$ is equal to the sum of the area of its orthogonal projection in the plane $\langle u,n\rangle=0$ multiplied by $2a$ and the area of the illuminated part in the direction n multiplied by b. The surface B is determined uniquely, to within a parallel transfer.

Theorem 2.40. In the unit sphere ω in R^3, a regular (k times continuously differentiable, $k\geq 7$, or analytic) function $\psi(n)$ is given such that

1) $n\psi(n)d\omega_n=0$,
2) $\rho(n)\leq 0$, $n\in\omega$,
3) $\rho(n)-\rho''_{ss}(n)\leq 0$, $n\in\omega$,

where

$$\rho(n)=\frac{1}{8\pi^2}\frac{d}{dt}\int\limits_{\langle n,p\rangle^2>t}\frac{(\Delta+2)\psi_1(p)\,|\langle n,p\rangle|}{\sqrt{\langle n,p\rangle^2-t}}d\omega_p\Bigg|_{t=0}+\frac{1}{4\pi^2 b}\int\limits_{\omega(n)}\frac{\Delta\psi^-(p)}{\langle n,p\rangle}d\omega_p.$$

$$\psi_1(p)=\frac{1}{a}\psi^+(p)-\frac{b}{4\pi a(a+b)}\int\limits_{\omega}\psi(p)d\omega_p,$$

The index s denotes differentiation in the point $n\in\omega$ along the arc s of any great circle. Then there exists a unique, to within a parallel transfer, regular (of the class $C^{k-3,\alpha}$, $0<\alpha<1$, or respectively, analytic) closed convex surface B, for which $\psi(n)$ is the sum of the area of its orthogonal projection in the plane $\langle x,n\rangle=0$ multiplied by $2a$ and the mean integral curvature of its illuminated part in the direction n multiplied by $2b$. The surface B is determined uniquely, to within a parallel transfer.

Proof of theorems 2.38-2.40 may be done by inverting the integral equation (theorem 2.33) and using the theorems of existence of convex surfaces with the given function of principal radii of curvature.

Stability. The stability theorems for determining convex surfaces by the functionals of their orthogonal projections and illuminated parts, which will be formulated below, are of the double interest. From the geometrical point of view, the estimates of proximity of surfaces which have similar functionals are very important. On the other hand, some of the functionals arise when considering the inverse scattering problem, taken in a certain approximation. In this connection, the stability theorems allow us to state the stability of determination of the scattering body from the scattering data, which, evidently, is of great importance in applications.

First, we shall give the stability esimate in determining a regular convex surface by the function $\varphi(n) = 2aF(n) + bS(n)$. Note, that the surface function $G(B, \cdot)$ of a regular convex surface B is expressed as the product of principle radii of curvature $R_1(n)R_2(n)$ by the formula (Pogorelov, 1969)

$$G(B, E) = \int_E R_1(n)R_2(n)d\omega_n, \quad E \subset \omega.$$

Further, using the stability estimate for the solution to the integral equation

$$\varphi(n) = a \int_\omega |\langle n, p\rangle|\, R_1(p)R_2(p)d\omega_p + b \int_{\omega(n)} R_1(p)R_2(p)d\omega_p$$

(Theorem 2.26) and applying the Volkov theorem of stability to solving the Minkovskii problem (Volkov, 1963), we obtain the following result.

Theorem 2.41. If two closed strictly convex regular surfaces B_1 and B_2 have the functions

$$\varphi_1(n) = 2aF_1(n) + bS_1(n),$$

$$\varphi_2(n) = 2aF_2(n) + bS_2(n)$$

being k times continuously differentiable, $k \geq 4$, and the inequality

$$\left\|\varphi_2(n) - \varphi_1(n)\right\|_{C^4(\omega)} \leq \varepsilon, \quad \varepsilon > 0$$

holds, then for a certain disposition of B_1 and B_2 the difference $H_2(n) - H_1(n)$ of their support functions satisfies the inequality

$$\left\|H_2(n) - H_1(n)\right\|_{C(\omega)} \leq c \cdot \varepsilon^{1/5},$$

for a certain constant $c > 0$.

To obtain the stability estimate in determining a regular convex surface by the function $\psi(n) = 2aL(n) + 2bM(n)$, we shall use the stability estimate for solving the Hristoffel' problem (Pogorelov, 1969): if $\rho_1(n) + \rho_2(n)$ ($\rho_2(n)$, $i = 1,2$ are the principal curvature radii of closed convex surfaces B_i) differ no more than on ε, then for a certain disposition of surfaces, their support functions $H_1(n)$, H_2 differ no more than on $(\ln 2 - 1/4)\varepsilon$.

This and the stability theorem for the integral equation

$$\psi(n) = a \int\limits_{\omega} |\langle n,p\rangle| \, [R_1(p) + R_2(p)] \, d\omega_p + b \int\limits_{\omega(n)} [R_1(p) + R_2(p)] \, d\omega_p$$

yield the following theorem.

Theorem 2.42. If two closed strictly convex regular surfaces B_1 and B_2 have k times continuously differentiable functions

$$\psi_1(n) = 2aL_1(n) + 2bM_1(n),$$

$$\psi_2(n) = 2aL_2(n) + 2bM_2(n),$$

and the inequality

$$\left\|\psi_2(n) - \psi_1(n)\right\|_{C^4(\omega)} \leq \varepsilon, \quad \varepsilon > 0$$

holds, then for a certain disposition of B_1 and B_2, we have

$$\left\|H_2(n) - H_1(n)\right\|_{C(\omega)} \leq c(\ln 2 - 1/4)\varepsilon,$$

where $c > 0$ is a constant.

Below we shall formulate some stability theorems which state that if the functionals of projections and illuminated parts of a convex surface are characterized by properties close to the properties of the functionals of projections and illuminated parts of the sphere, then this surface is close to the sphere.

Theorem 2.43. For a closed strictly convex regular surface B, let the function $\varphi(n) = 2aF(n) + bS(n)$ satisfy the inequality

$$\left|\frac{1}{8\pi^2}\frac{d}{dt}\int\limits_{\langle n,p\rangle^2 > t} \frac{(\Delta + 2)\varphi_1(p)\langle n,p\rangle}{\sqrt{\langle n,p\rangle^2 - t}} d\omega_p + \frac{1}{4\pi^2 b}\int\limits_{\omega(n)} \frac{\Delta\varphi^-(p)}{\langle n,p\rangle} d\omega_p + 1\right| \leq \varepsilon,$$

where $n \in \omega$. Then the surface B is contained in the ball of radius $1 + c\sqrt{\varepsilon}$ and contains the ball of radius $1 - c\sqrt{\varepsilon}$; $c > 0$ is a constant.

Let R and r be the radii of the circumscribed and inscribed balls for a convex surface B, and $R(n)$, $r(n)$ be the radii of the circumscribed and inscribed disks of the orthogonal projections of B in the plane $\langle x, n\rangle = 0$, $n \in \omega$.

Theorem 2.44. If $R^2(n) - r^2(n) \leq \varepsilon^2$, then $R - r \leq 2\varepsilon$.

Bearing in mind theorem 2.43, it is natural to expect that if the isoperimetric defect $L^2(n) - 4\pi F(n)$ is small for each orthogonal projection $B(n)$ of B in the plane $\langle x, n\rangle = 0$, then B is close to the sphere.

Really, the following theorem holds.

Theorem 2.45. If $L^2(n) - 4\pi F(n) \leq \varepsilon^2$, then $R - r < \varepsilon/\pi$.

Determination of nonconvex manifolds. In this section we shall consider the geometric problems of determining nonconvex manifolds by convex hull and orthogonal projections. The theorems from this section are contained in Anikonov and Golubyatnikov (1978), and in Golubyatnikov (1982). First, we shall give the theorems connected with convex hulls.

Theorem 2.46. An analytic connected compact k-dimensional manifold without boundary analytically imbedded in R^m, $m \geq 3$ is uniquely determined by its convex hull.

Theorem 2.47. If a convex hull of a k-dimensional connected compact manifold without boundary in R^m has an analytic boundary, then the external manifold boundary is a convex analytic surface.

Applying these results leads to unique determination of analytic and other nonconvex manifolds by their projections and functionals.

Let B be a compact connected manifold, analytic, centrally symmetric, and imbedded in R^m. Denote by $h(n)$ the width of B in the direction $n \in \omega = \{|x| \in R^m, \ |x| = 1\}$.

Theorem 2.48. The manifold B is determined uniquely, to within a parallel transfer by $h(n)$, $n \in \omega$.

Let $F(n)$ be the area of the convex hull of the manifold B projection (not necessary centrally symmetric) in the plane $\langle x, n\rangle = 0$; $S(n)$ be the area of the illuminated part of the convex hull of B.

Theorem 2.49. The manifold B is uniquely determined by the function $\varphi(n) = 2aF(n) + bS * n$, $a \neq 0$, $b \neq 0$, $a + b \neq 0$, $n \in \omega$.

In the case when B is a centrally symmetric manifold, the following theorem holds.

Theorem 2.50. The manifold B is uniquely determined by the function $F(n)$, $n \in \omega$.

Golubyatnikov (1982) introduced the classes of sets for which the Zyuss theorem (on matching by a parallel transfer of two sets if their projections may be matched) holds. These are analytic manifolds and bounded visible sets. (A set in R^m is called visible if for each point which does not belong to this set, there exists a line which does not intersect this set). Theorem 2.46 and the Zyuss theorem yield the following theorems.

Theorem 2.51. (Golubyatnikov, 1982). If for two analytic closed surfaces, their orthogonal projections in the plane $\langle x, n\rangle = 0$, $n \in \omega$ may be matched by a parallel transfer in this plane, then these surfaces may be matched by a parallel transfer in space.

Theorem 2.52. (Golubyatnikov, 1982). If the projections of two visible compact sets in R^m may be matched by parallel transforms, then these compact sets may be matched by a parallel transform.

Note, in conclusion, that probably, the following statement also holds.

A connected compact without boundary analytic manifold $B \subset R^m$, $m \geq 2$ is uniquely determined if one of the functions

$$\varphi(n) = aF(n) + bS(n),$$

$$\psi(n) = aL(n) + bM(n), \quad n \in \omega$$

is known. Here $F(n)$, $S(n)$ are the areas of the orthogonal projection and the illuminated part of B respectively; $L(n)$ and $M(n)$ are the length of projection boundary and the mean integral curvature of the illuminated part of B respectively.

Chapter 3

Numerical solution of direct and inverse problems.

Here we shall consider the numerical solution of mathematical simulation of seismic and acoustic fields from complicated sources on the basis of measured information.

The peculiarity of these sources is their possible motion. These sources have been thoroughly investigated by Blokhintsev (1981). His monograph was first published in 1945. In our monograph the ideas of Blokhintsev have been developed, especially, in connection with application to seismic fields.

As for solution of inverse problems, because of their instability, the main focus of attention has been the construction of optimal methods for inversion of regularization.

§3.1 Mathematical simulation on the basis of the seismic moment tensor theory.

In this paragraph we shall deal with the problems of simulation of a seismic vibration field from sources which are characterized by a wide spectrum of radiation, by complicated spatial directedness, and by their spatial motion.

The field from a fixed source. Let the shift vector $u(x,t)$ satisfy the motion equation of the homogeneous ideal elastic medium

$$\frac{\partial \sigma_{ij}}{\partial x_j} - \rho \frac{\partial^2 u_i}{\partial t^2} = f_i(x,t) \tag{1.1}$$

where $x \in R^3$, $t \in R^1$; ρ is the medium density; $f(x,t)$ is the vector of side equivalent forces density; σ_{ij} is the tensor connected with the shift vector $u(x,t)$ as follows

$$\sigma_{ij} = \mu \left(\frac{\partial u_i}{\partial x_j} + \frac{\partial u_j}{\partial x_i} \right) + \lambda \sigma_{ij} \frac{\partial u_k}{\partial x_k};$$

μ, λ are the Lamé constants; the repetition of indices denotes summing.

From the practical point of view the equivalent forces are of an interest. They describe a shift gap, a shift blow-up, a transformed volume change and so on. In Aki and Richards (1983), and in Rays (1982) it was shown that these forces are as follows:

$$f_\beta(x,t) = -\frac{\partial m_{\alpha\beta}(x,t)}{\partial x_\alpha}, \tag{1.2}$$

where $m_{\alpha\beta}(x,t)$ is the so-called tensor of seismic moment density. Further, as in these monographs, we shall assume that this tensor is symmetric. Let

$$m_{\alpha\beta}(x,t) = M_{\alpha\beta}(t)\delta(x_1 - y_1)\delta(x_2 - y_2)\delta(x_3 - y_3). \tag{1.3}$$

Thus, we consider the effective point mathematical model of a source.

Here $M_{\alpha\beta}(t)$ is the seismic moment tensor; x_i, y_i are the coordinates of receiver and source respectively. The form of $M_{\alpha\beta}(t)$ influences the type of modelling of seismic source. So, for example,

$$M_{\alpha\beta}(t) = \begin{pmatrix} 0 & 0 & \mu U_{av} S \\ 0 & 0 & 0 \\ \mu U_{av} S & 0 & 0 \end{pmatrix} \tag{1.4}$$

effectively determines a point source of the shift type in the plane $x_3 = 0$ along the axis x_1 (or conversely), i.e., the double dipole without a moment. Here $U_{av}(t)$ has the sense of an average shift; $S(t)$ is a certain effective area element in the moment t.

The tensor

$$M_{\alpha\beta}(t) = \begin{pmatrix} 0 & 0 & 2\mu V \Delta e_{13} \\ 0 & 0 & 0 \\ 2\mu V \Delta e_{13} & 0 & 0 \end{pmatrix} \tag{1.5}$$

has the structure of the (3.3) tensor and corresponds to the analogous source, but physically describes another mechanism of wave formation, namely, the rapid shift destruction of the body (Aki and Richards, 1983). In (3.5), $\Delta e_{13} = \Delta e_{31}$ are the components of transformed deformations of the body V.

The tensor

$$M_{\alpha\beta}(t) = \begin{pmatrix} \lambda U_{av} S & 0 & 0 \\ 0 & \lambda U_{av} S & 0 \\ 0 & 0 & (\lambda + 2\mu) U_{av} S \end{pmatrix}$$

describes the centre of crack separation type, and the seismic moment tensor

$$M_{\alpha\beta}(t) = \frac{4\pi}{3} a^3 \begin{pmatrix} \Delta p & 0 & 0 \\ 0 & \Delta p & 0 \\ 0 & 0 & \Delta p \end{pmatrix}$$

corresponds to transformation increase of spherical volume of the radius a. Here $\Delta p = \left(\lambda + \frac{2}{3}\mu\right) \delta\Theta$; $\Delta\Theta$ is the relative volume change (Aki and Richards, 1983).

The solution to (3.1) with right side (3.2), (3.3) is constructed by means of the Green function for free space

$$u_\nu(x,t) = \int_{-\infty}^{t} \int_{R^3} G_{\nu\beta}(x, x', t - t') f_\beta(x', t') dx' dt'.$$

It is represented as the sum of two terms describing propagation of longitudinal and transverse seismic waves (Rays, 1982):

$$\hat{u}_\nu(x,\omega) = \hat{u}^p_\nu(x,\omega) + \hat{u}^s_\nu(x,\omega), \tag{1.6}$$

$$\hat{u}^p_\nu(x,\omega) = \frac{\partial^3 \hat{\mu}_{\alpha\beta}(x,\omega;c_p)}{\partial x_\alpha \partial x_\beta \partial x_\nu}, \tag{1.7}$$

$$\hat{u}^s_\nu(x,\omega) = \Delta \frac{\partial}{\partial x_\beta}\hat{\mu}_{\alpha\beta}(x,\omega;c_s) - \frac{\partial^3 \hat{\mu}_{\alpha\beta}(x,\omega;c_s)}{\partial x_\alpha \partial x_\beta \partial x_\nu}. \tag{1.8}$$

Here Δ is the Laplace operator; $\hat{u}^p_\nu(x,\omega)$, $\hat{u}^s_\nu(x,\omega)$ are the Fourier transforms of longitudinal and transverse components of elastic waves; c_p, c_s are the speeds of propagation of these waves ($c_s^2 = \mu/\rho$, $c_p = (\lambda + 2\mu)/\rho$). The function $\hat{\mu}_{\alpha\beta}$ is as follows:

$$\hat{\mu}_{\alpha\beta}(x,\omega;c) = \hat{M}_{\alpha\beta}(\omega)\frac{\exp\left(-i\frac{\omega}{c}R\right)}{4\pi\rho\omega^2 R}\,, \tag{1.9}$$

where $R = \sqrt{(x_1-y_1)^2 + (x_2-y_2)^2 + (x_3-y_3)^2}$ and $\hat{M}_{\alpha\beta}(\omega)$ is the Fourier transformation by time of the seismic moment tensor.

Substituting (3.9) into (3.7), (3.8) and differentiating by the spatial variables, we can represent formulas (3.7), (3.8) as follows

$$\hat{u}^p_\nu(x,\omega) = \frac{\exp\left(-i\frac{\omega}{c_p}R\right)}{4\pi\rho}\hat{M}_{\alpha\beta}(\omega)$$

$$\times\left(\frac{-15\gamma_\alpha\gamma_\beta\gamma_\nu + 3\gamma_\alpha\delta_{\beta\nu} + 3\gamma_\beta\delta_{\alpha\nu} + 3\gamma_\nu\delta_{\alpha\beta}}{\omega^2 R^4}\right.$$

$$-\frac{15\gamma_\alpha\gamma_\beta\gamma_\nu - 3\gamma_\alpha\delta_{\beta\nu} - 3\gamma_\beta\delta_{\alpha\nu} - 3\gamma_\nu\delta_{\alpha\beta}}{\omega c_p R^3}i$$

$$\left.-\frac{6\gamma_\alpha\gamma_\beta\gamma_\nu + \gamma_\alpha\delta_{\beta\nu} + \gamma_\beta\delta_{\alpha\nu} + \gamma_\nu\delta_{\alpha\beta}}{c_p^2 R^2} + i\frac{\omega}{Rc_p^3}\gamma_\alpha\gamma_\beta\gamma_\nu\right); \tag{1.10}$$

$$\hat{u}^s_\nu(x,\omega) = \frac{\exp\left(-i\frac{\omega}{c_s}R\right)}{4\pi\rho}\left[\hat{M}_{\nu\beta}(\omega)\left(\frac{-15\gamma_\nu^2\gamma_\beta + 6\gamma_\nu\delta_{\beta\nu} + 3\gamma_\beta}{\omega^2 R^4}\right.\right.$$

$$-\frac{15\gamma_\nu^2\gamma_\beta - 6\gamma_\nu\delta_{\beta\nu} - 3\gamma_\beta}{c_s\omega R^3}i - \frac{6\gamma_\nu^2\gamma_\beta + 2\gamma_\nu\delta_{\beta\nu} + \gamma_\beta}{c_s^2 R^2}$$

$$+ \frac{\omega}{c_s^3 R} i\gamma_\nu^2\gamma_\beta\Big) - \hat{M}_{\alpha\beta}(\omega)\Bigg(\frac{-15\gamma_\alpha\gamma_\beta\gamma_\nu + 3\gamma_\alpha\delta_{\beta\nu} + 3\gamma_\beta\delta_{\alpha\nu} + 3\gamma_\nu\delta_{\alpha\beta}}{\omega^2 R^4}$$

$$-\frac{-15\gamma_\alpha\gamma_\beta\gamma_\nu - 3\gamma_\alpha\delta_{\beta\nu} - 3\gamma_\beta\delta_{\alpha\nu} - 3\gamma_\nu\delta_{\alpha\beta}}{\omega c_s R^3} i$$

$$-\frac{-6\gamma_\alpha\gamma_\beta\gamma_\nu + \gamma_\alpha\delta_{\beta\nu} + \gamma_\beta\delta_{\alpha\nu} + \gamma_\nu\delta_{\alpha\beta}}{c_s^2 R^2} + i\frac{\omega}{R c_s^3}\gamma_\alpha\gamma_\beta\gamma_\nu\Bigg)\Bigg], \tag{1.11}$$

where $\gamma_j = (x_j - y_j)/R$; $\delta_{\alpha\beta}$ is the Kronecker symbol, $i^2 = -1$.

On the basis of these formulas, a program to calculate the signal spectrum was written in FORTRAN language. In the Figures 3.1, 3.2 we may see the levels of the amplitude spectrum of the signal for longitudinal and transverse waves. The source position is $y = (0, 0, 10 \text{ km})$; the type is the double dipole without moment of the form (3.4)-(3.5)

$$M_{\alpha\beta}(t) = \delta(t) = \begin{pmatrix} 0 & 0 & b \\ 0 & 0 & 0 \\ b & 0 & 0 \end{pmatrix}, \tag{1.12}$$

where $b = 10^{12}\text{J} \cdot \text{s}$. Here and below $\omega = 1\text{Hertz}$.

Thus, in the physical sense, the source describes either the instantaneous gap in the plane $x_3 = 0$ (source of the form (3.4)), or the instantaneous destruction of a certain volume (source of the form (3.5)). Parameters of the medium here and below are $c_p = 6.1$ km/s, $c_s = 3.5$ km/s, $\rho = 2.72$ g/cm^3.

Figure 3.3 represents the spectral distributions for the signal phase $\text{Im}(\hat{u}) = (\hat{u})\sin\varphi$, where $\varphi = \arg(u) = \text{arctg}(\text{Im}u/\text{Re}u)$. Evidently, such signal representation has information both on shift in the spatial point and on its direction, which is important for interpreting the results of the simulation.

In Figure 3.4 we see the shift field for an arbitrary chosen seismic moment tensor

$$M_{ij}(t) = \delta(t) \cdot a \cdot \begin{pmatrix} 1 & 4 & 5 \\ 4 & 2 & 6 \\ 5 & 6 & 3 \end{pmatrix}, \tag{1.13}$$

where $a = 10^{12}\text{J} \cdot \text{s}$.

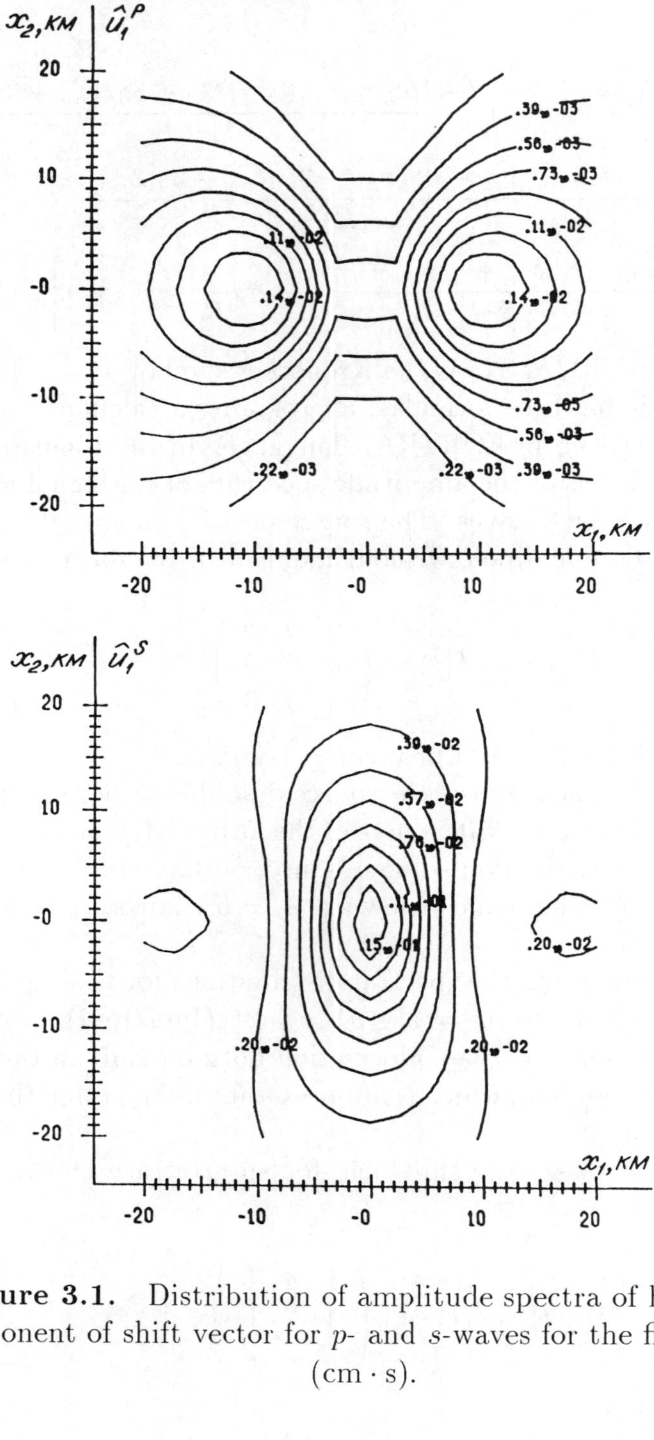

Figure 3.1. Distribution of amplitude spectra of horizontal component of shift vector for p- and s-waves for the fixed source $(\mathrm{cm} \cdot \mathrm{s})$.

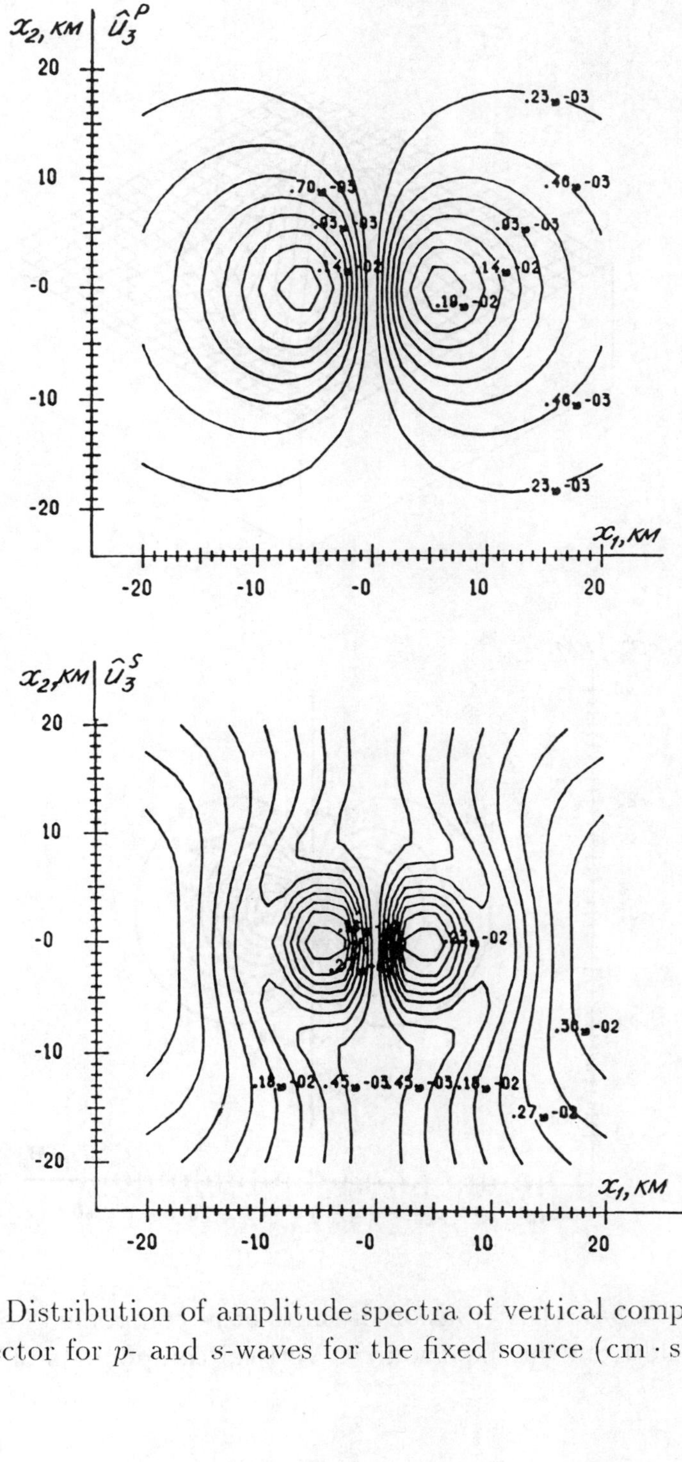

Figure 3.2. Distribution of amplitude spectra of vertical component of shift vector for p- and s-waves for the fixed source (cm · s).

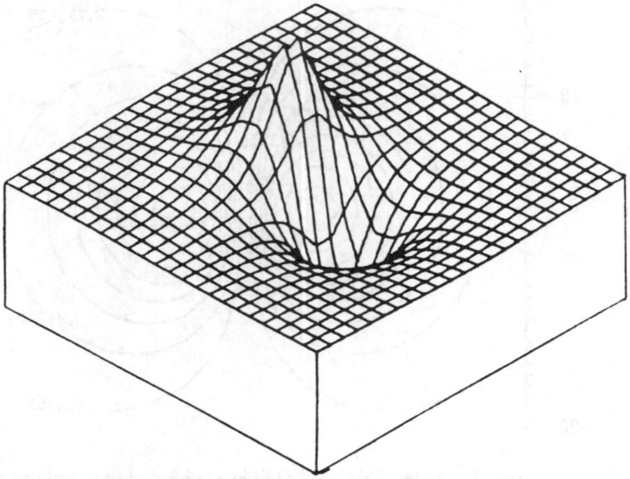

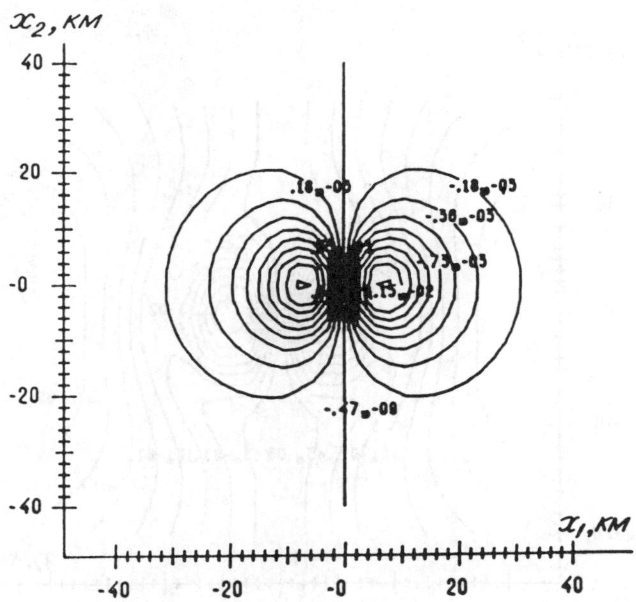

Figure 3.3a. Distribution of shifts of p- components of the seismic field from the fixed source. Vertical components of p-wave.

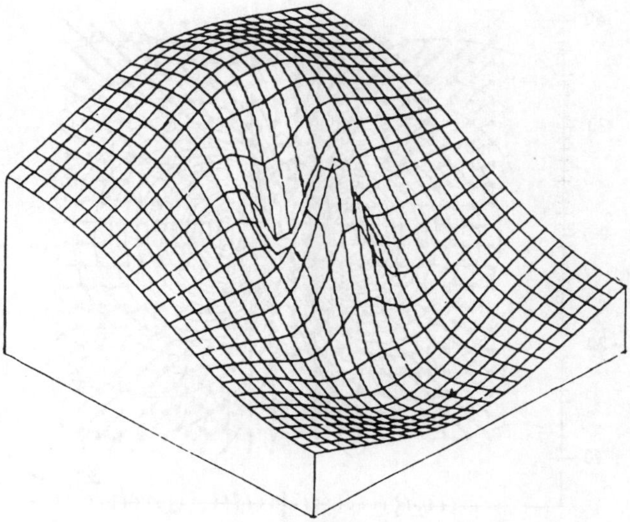

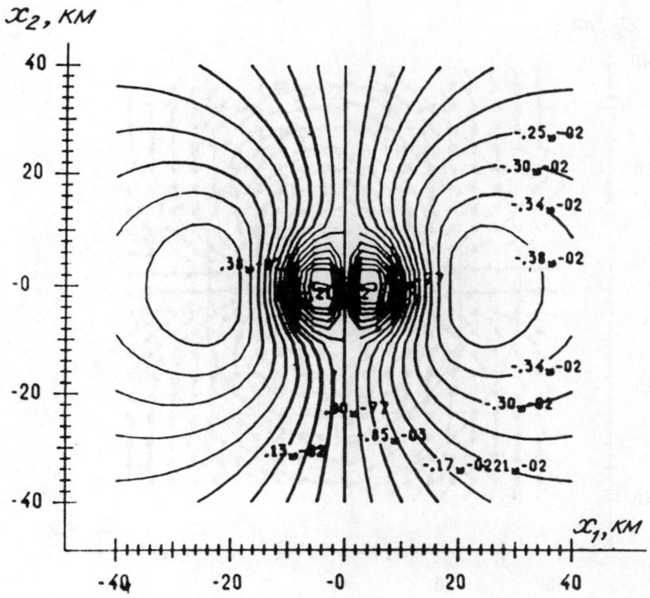

Figure 3.3b. Distribution of shifts of s- components of the seismic field from the fixed source. Vertical components of s-wave.

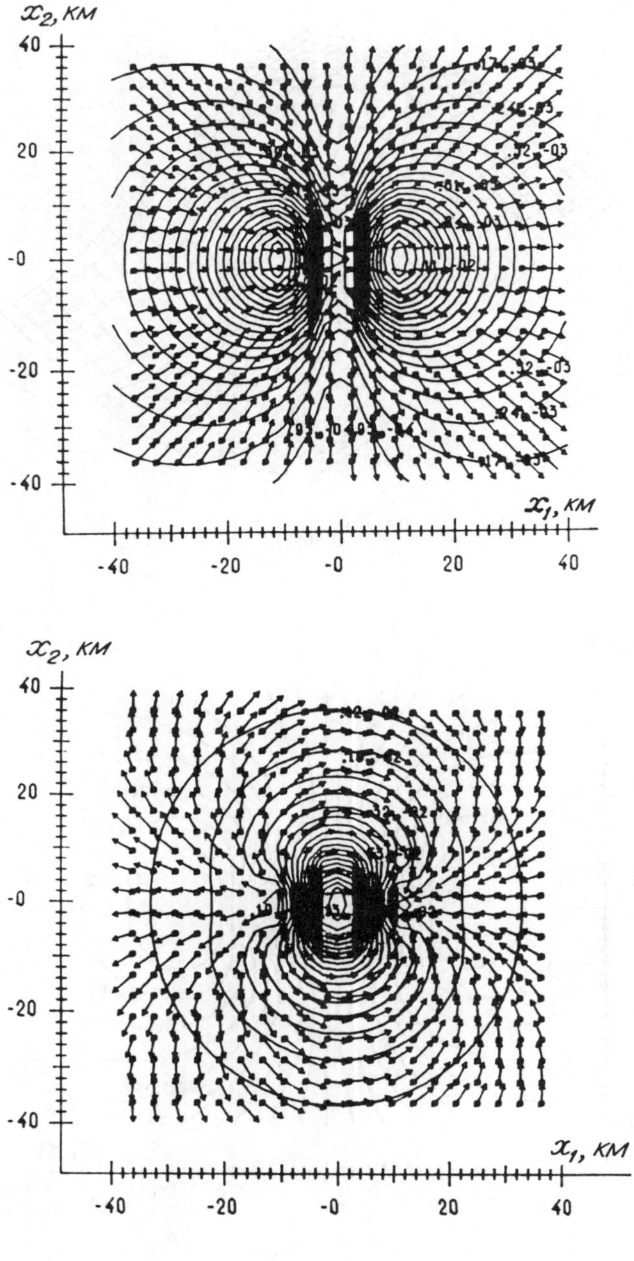

Figure 3.3c. Distribution of shifts of p- components of the seismic field from the fixed source. Horizontal components of p- wave.

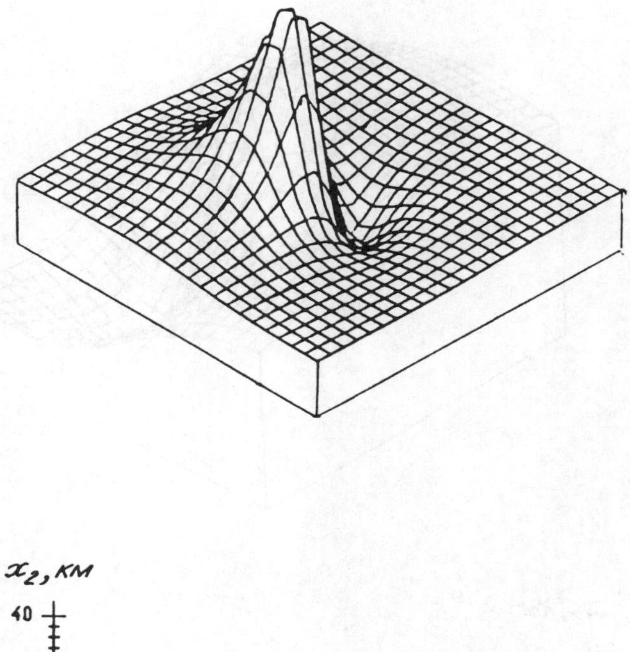

Figure 3.4a. Distribution of shifts of p- components of the seismic field from the fixed source for the tensor of the form (3.13). Vertical components of p-wave.

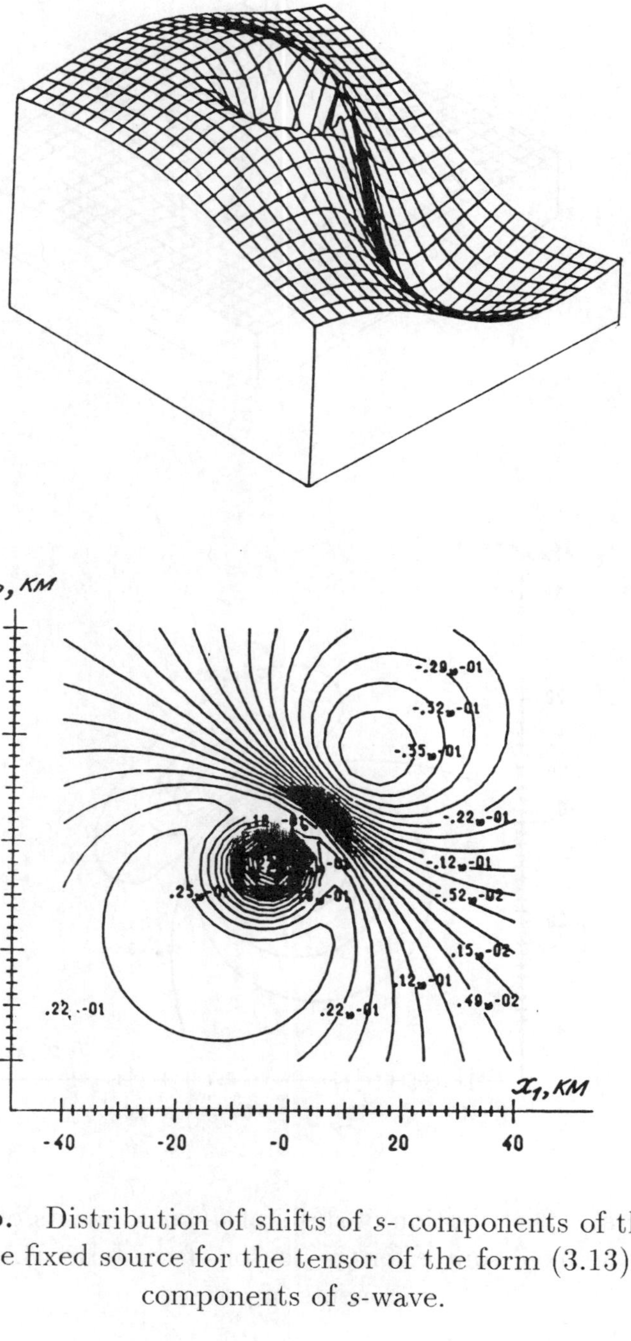

Figure 3.4b. Distribution of shifts of s- components of the seismic field from the fixed source for the tensor of the form (3.13). Vertical components of s-wave.

The moving source. Solving in the frequency domain. Let the right side of the Lamé system (3.1) be

$$f_i(x,t) = \frac{\partial}{\partial x_j} M_{ij}(t)\delta(x - vt - y), \tag{1.14}$$

where $M_{ij}(t)$ is the seismic moment tensor; $v = (v_1, v_2, v_3)$ is the speed vector ($|v| < c_s$); $y = (y_1, y_2, y_3)$ is the vector of source disposition in the moment $t = 0$.

Solution to (3.1) with right side (3.14) in the case of moving sources in terms of the theoretical instantaneous spectrum is represented as

$$u_\nu(x,t) = u_\nu^p(x,t) + u_\nu^s(x,t); \tag{1.15}$$

$$u_\nu^p(x,t) = U_\nu^p(x,t,t_0)\Big|_{t=t_0}; \quad u_\nu^s(x,t) = U_\nu^s(x,t,t_0)\Big|_{t=t_0}; \tag{1.16}$$

$$U_\nu^p(x,t,t_0) = \int\limits_{R^1} \hat{U}_\nu^p(x,\omega,t_0)e^{i\omega t}d\omega; \quad U_\nu^s(x,t,t_0) = \int\limits_{R^1} \hat{U}_\nu^s(x,\omega,t_0)e^{i\omega t}d\omega; \tag{1.17}$$

$$U_\nu^p(x,\omega,t_0) = \frac{\partial^3 \hat{\mu}_{\alpha\beta}(x,\omega,t_0,c_p)}{\partial x_\alpha \partial x_\beta \partial x_\nu}; \tag{1.18}$$

$$U_\nu^s(x,\omega,t_0) = \Delta \frac{\partial}{\partial x_\beta} \hat{\mu}_{\alpha\beta}(x,\omega,t_0,c_s) - \frac{\partial^3 \hat{\mu}_{\alpha\beta}(x,\omega,t_0,c_s)}{\partial x_\alpha \partial x_\beta \partial x_\nu}. \tag{1.19}$$

Here $U_\nu^p(x,\omega,t_0)$, $\hat{U}_\nu^s(x,\omega,t_0)$ are the instantaneous spectra of ν-components of longitudinal and transverse waves of the signal registered in the point (x_1, x_2, x_3) in the moment when the source was in the point $x_{\text{source}} = vt_0 + y$. Formulas (3.15), (3.18), (3.19) are analogous to the formulas for the fixed source (see (3.6)-(3.8)). The functions $\hat{\mu}_{\alpha\beta}(x,\omega,t_0,c)$, $c = c_p$, $c = c_s$ are as follows

$$\hat{\mu}_{\alpha\beta}(x,\omega,t_0,c) = \hat{M}_{\alpha\beta}(\omega)\frac{\exp\left(-i\frac{\omega}{c}R^*\right)}{4\pi\rho\omega^2 R}, \tag{1.20}$$

where $\hat{M}_{\alpha\beta}(\omega)$ is the Fourier transform of $M_{\alpha\beta}(t)$, and R, R^* are determined as

$$R(x, t_0, v, y) = \left((r, v)^2 + (r, r)c^2\right)^{1/2} / c, \tag{1.21}$$

$$R^*(x, t_0, v, y) = (r, v)/c + R, \tag{1.22}$$

$$r = x - vt - y.$$

Instantaneous spectra (3.18), (3.19) are determined by the formulas

$$\hat{U}_\nu^p(x, \omega, t_0) = \hat{M}_{\alpha\beta}(\omega) \frac{\exp\left(i \frac{\omega}{c_p} R^*\right)}{4\pi \rho R}$$

$$\times \left(i \frac{\omega}{c_p^3} A^{\alpha\beta\nu} + \frac{1}{c_p^2} B^{\alpha\beta\nu} + \frac{1}{c_p \omega} C^{\alpha\beta\nu} + \frac{1}{\omega^2} D^{\alpha\beta\nu} \right);$$

$$\hat{U}_\nu^s(x, \omega, t_0) = \frac{\exp\left(i \frac{\omega}{c_s} R^*\right)}{4\pi \rho R} \left[\hat{M}_{\alpha\beta}(\omega) \left(i \frac{\omega}{c_s^3} A^{\nu\beta\nu} + \frac{1}{c_s^2} B^{\nu\beta\nu} + \frac{1}{c_s \omega} C^{\nu\beta\nu} \right.\right.$$

$$\left.\left. + \frac{1}{\omega^2} D^{\nu\beta\nu} \right) - \hat{M}_{\alpha\beta}(\omega) \left(i \frac{\omega}{c_s^3} A^{\alpha\beta\nu} + \frac{1}{c_s^2} B^{\alpha\beta\nu} + \frac{1}{c_s \omega} C^{\alpha\beta\nu} + \frac{1}{\omega^2} D^{\alpha\beta\nu} \right) \right].$$

Here

$$A^{\alpha\beta\nu}(x, t_0, v, y, c) = \frac{\partial R^*}{\partial x_\alpha} \frac{\partial R^*}{\partial x_\beta} \frac{\partial R^*}{\partial x_\nu}.$$

In Figures 3.5-3.10 we may see the results of mathematical simulation of instantaneous spectra for a moving source with the seismic moment tensor (3.12). The source is moving along the x_1 axis with the speed 3 km/s ($v = (3, 0, 0)$). The centre depth, as earlier, is 10 km. The medium parameters are the same.

Analysis of the wave field shows the importance of the final speed of propagation from the centre for a complete picture of seismic process dynamics.

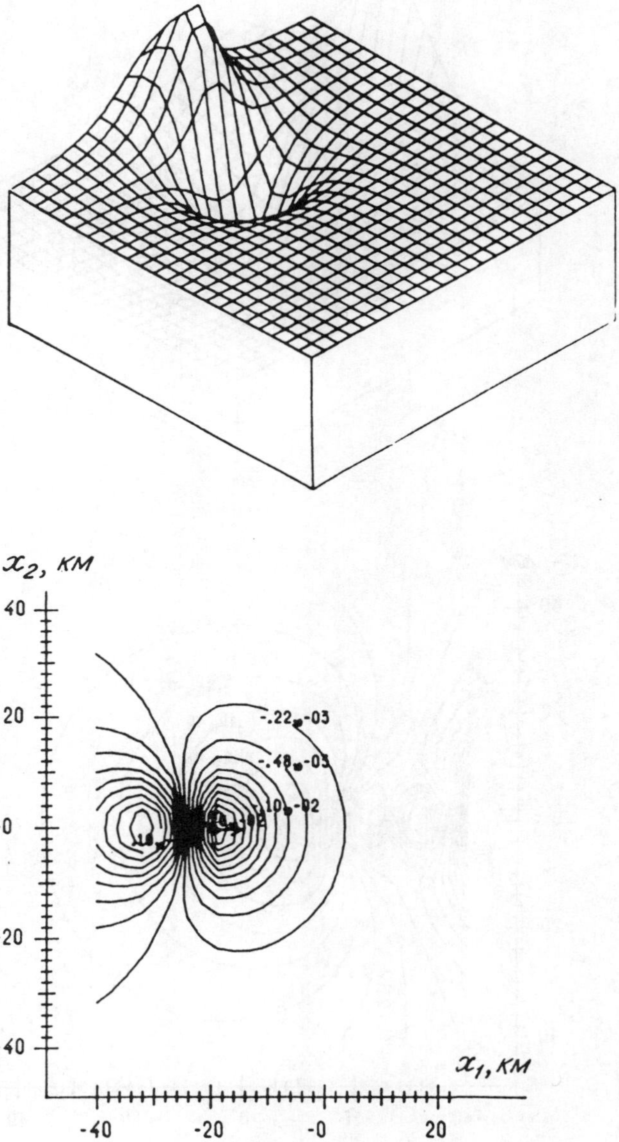

Figure 3.5a. The vertical component of instantaneous distribution of shift of p- components of a seismic field from a moving source in the moment $t = -8$ s, $v = 2.5$ km/s; the source coordinates (-20 km, 0.10 km).

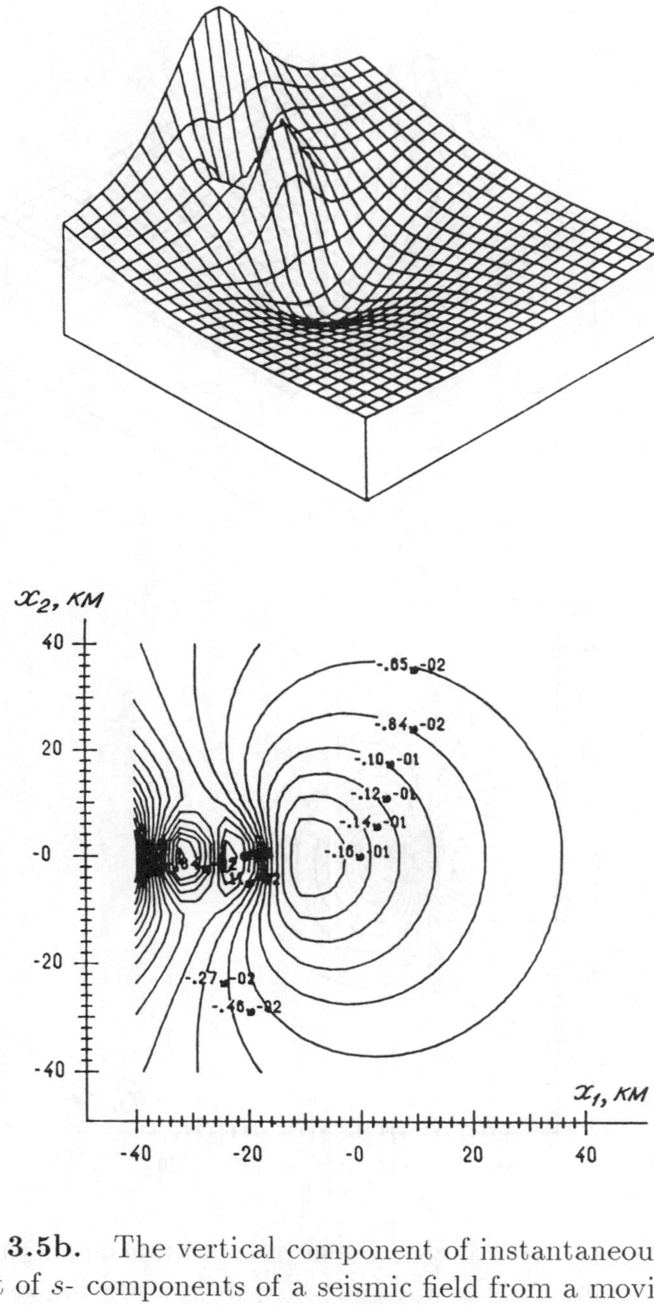

Figure 3.5b. The vertical component of instantaneous distribution of shift of s- components of a seismic field from a moving source in the moment $t = -8$ s, $v = 2.5$ km/s; the source coordinates $(-20 \text{ km}, 0.10 \text{ km})$.

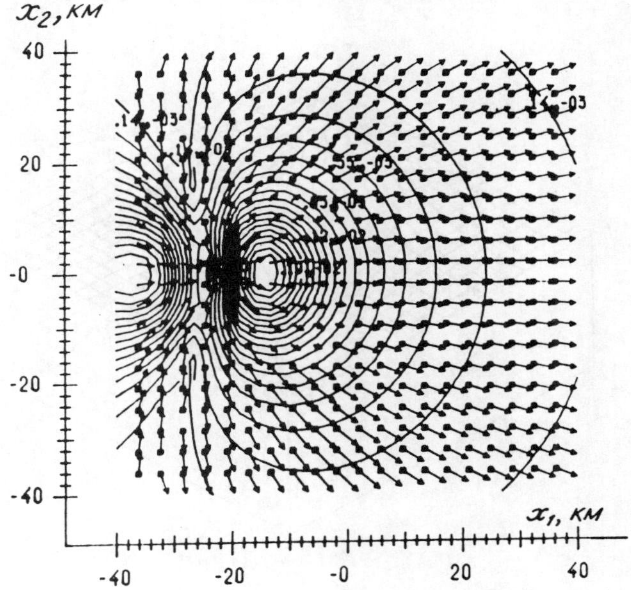

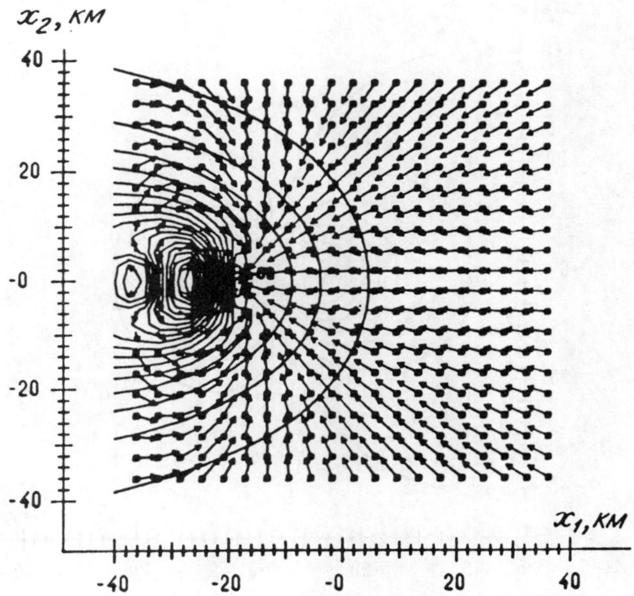

Figure 3.6. The horizontal component of instantaneous distribution of shift of p- and s- components of seismic field from the moving source in the moment $t = -8$ s, $v = 2,5$ km/s; the source coordinates $(-20 \text{ km}, 0.10 \text{ km})$. a - p-wave, b - s-wave.

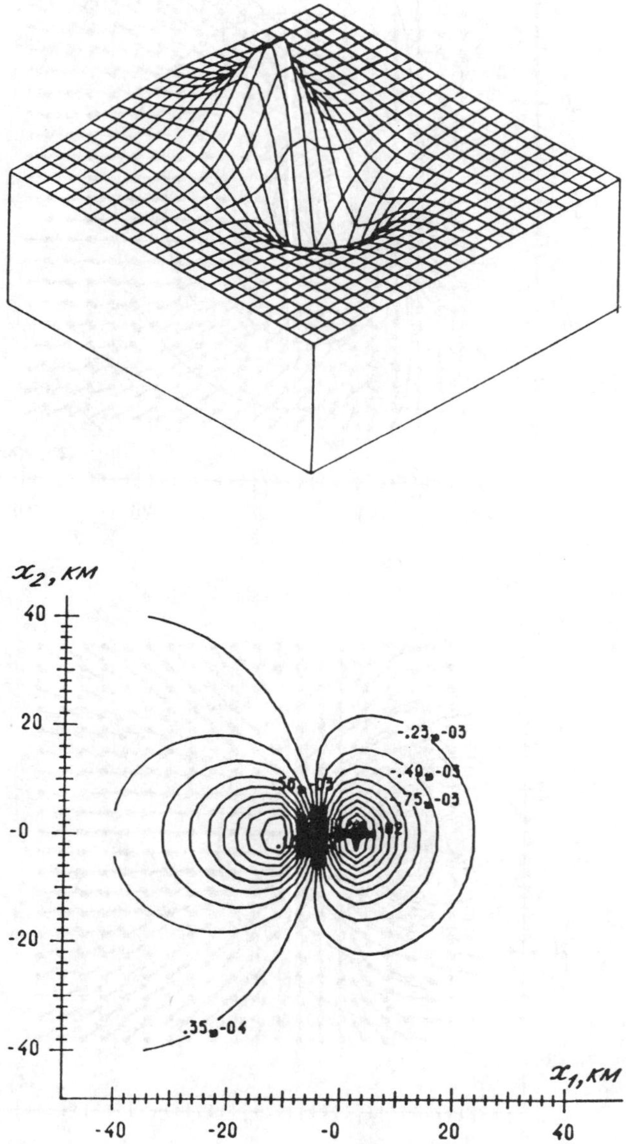

Figure 3.7. The instantaneous distribution of shifts of p- and s-components of a seismic field from a moving source in the moment $t = 0$, $v = 2.5$ km/s; the source coordinates $(0, 0.10$ km). The vertical component of p-wave.

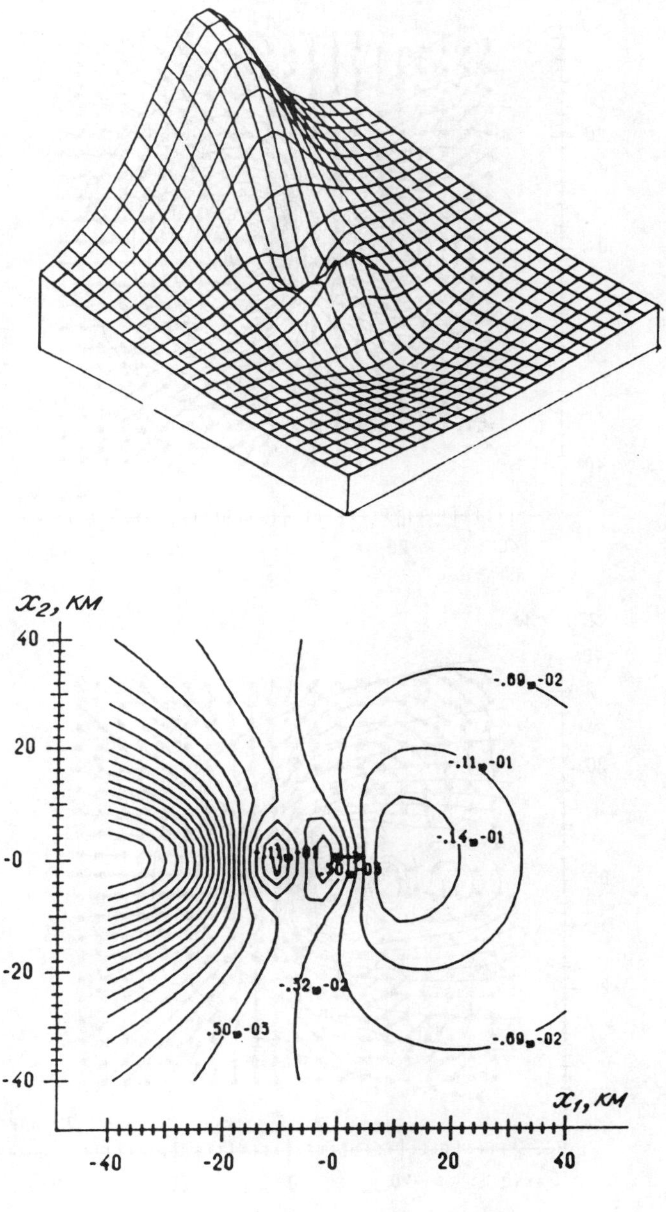

Figure 3.8. The instantaneous distribution of shifts of p- and s-components of a seismic field from a moving source in the moment $t = 0$, $v = 2.5$ km/s; the source coordinates $(0, 0.10$ km$)$. The vertical component of s-wave.

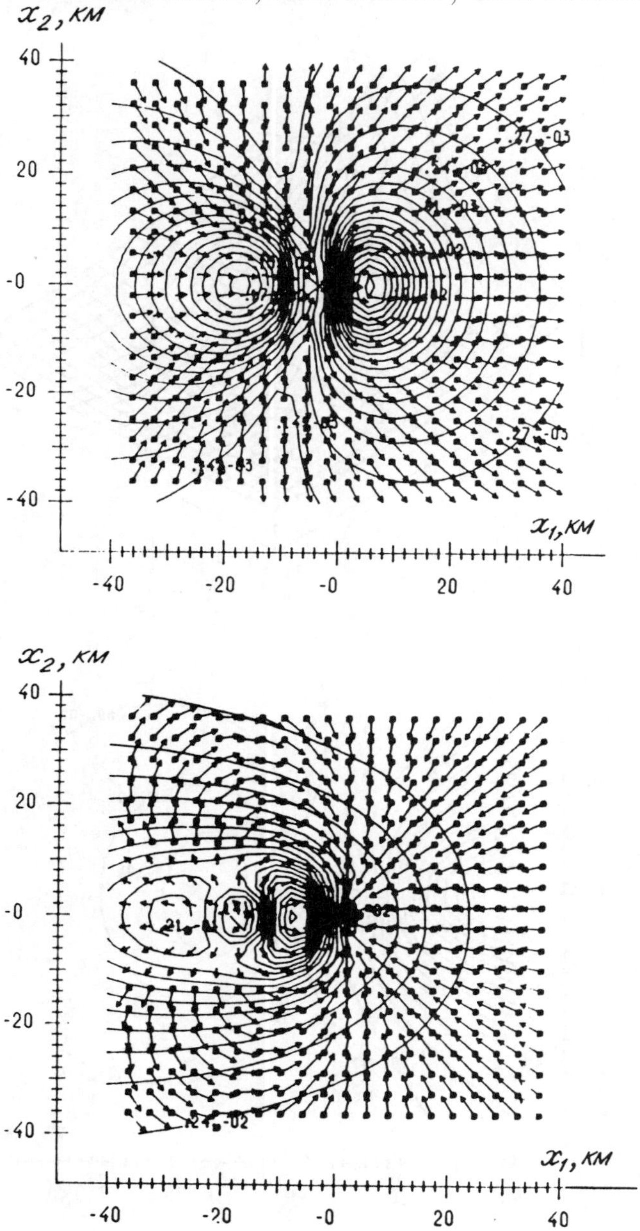

Figure 3.9. The instantaneous distribution of shifts of p- and s-components of a seismic field from a moving source in the moment $t = 0$, $v = 2.5$ km/s; the source coordinates $(0, 0.10$ km$)$. The horizontal components of p- and s-waves.

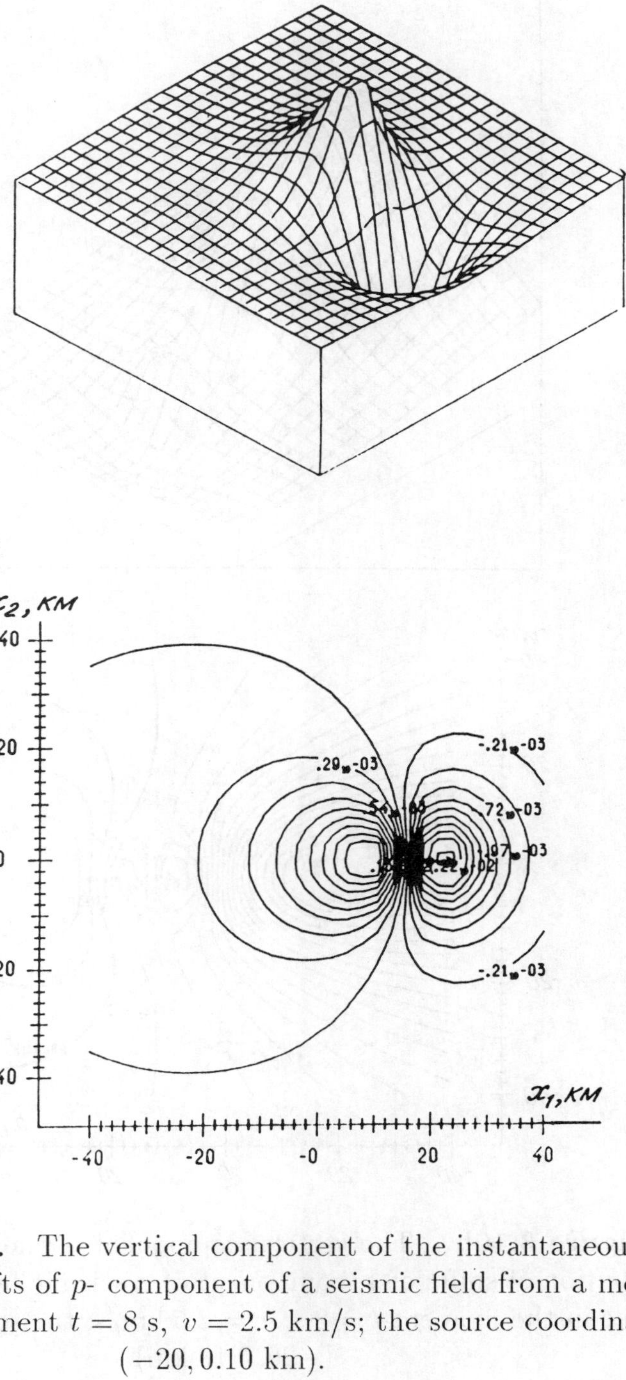

Figure 3.10a. The vertical component of the instantaneous distribution of shifts of p- component of a seismic field from a moving source in the moment $t = 8$ s, $v = 2.5$ km/s; the source coordinates $(-20, 0.10$ km$)$.

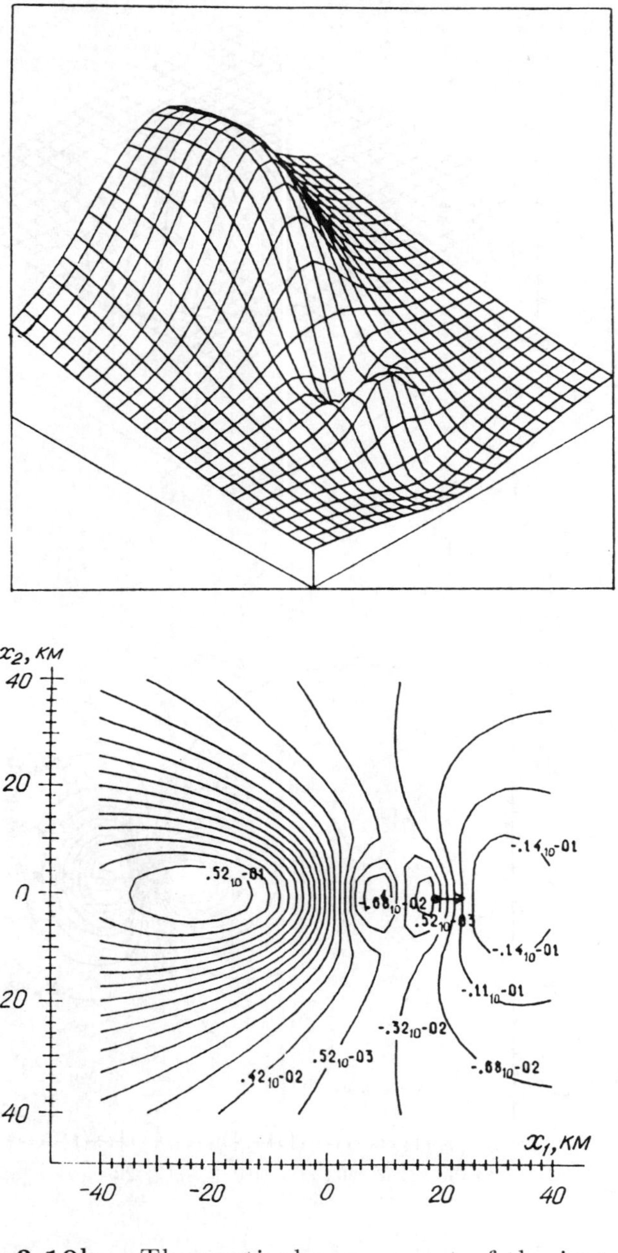

Figure 3.10b. The vertical component of the instantaneous distribution of shifts of s- component of a seismic field from a moving source in the moment $t = 8$ s, $v = 2.5$ km/s; the source coordinates $(-20, 0.10$ km).

Moving source. Solving in the time domain. To construct theoretical seismograms from moving sources (3.14) it is desirable to have explicit formulas for solution to the direct problem in the space-time domain. For this purpose we need to find the inverse Fourier transform for the function $\hat{\mu}_{\alpha\beta}(x,\omega,t_0,c)$ of the form (3.20)

$$\mu_{\alpha\beta}(x,t,c) = \int\limits_{R^1} \hat{\mu}_{\alpha\beta}(x,\omega,t,c)e^{i\omega t}d\omega.$$

It may be shown that this solution is as follows

$$\mu_{\alpha\beta}(x,t,c) = \frac{1}{4\pi\rho}\int\limits_{-\infty}^{t}\int\limits_{-\infty}^{\tau}\frac{M_{\alpha\beta}(P)}{R}dqd\tau, \qquad (1.23)$$

where

$$P(x,q;v,y,c) = q - \frac{R^*(x,q,v,y,c)}{c};$$

the functions R^*, R are determined as before (formulas (3.21)-(3.22)).

In this connection the components of longitudinal and transverse waves are determined by the formulas

$$u_\nu^p(x,t) = L_{c_p}^{\alpha\beta\nu}m_{\alpha\beta}(x,t),$$

$$u_\nu^s(x,t) = L_{c_s}^{\alpha\alpha\beta}m_{\nu\beta}(x,t) - L_{c_s}^{\alpha\beta\nu}m_{\alpha\beta}(x,t),$$

where the operator L_c acts on the tensor of seismic moment density $m_{\alpha\beta}(x,t)$ as follows:

$$L_c^{\alpha\beta\nu}m_{\alpha\beta}(x,t) = M_{\alpha\beta}(P)\Theta^{\alpha\beta\nu} + \dot{M}_{\alpha\beta}(P)Q^{\alpha\beta\nu}$$

$$+\int\limits_{-\infty}^{t} M_{\alpha\beta}\big(P(\tau)\big)S^{\alpha\beta\nu}(\tau)d\tau + \int\limits_{-\infty}^{t}\int\limits_{-\infty}^{\tau} M_{\alpha\beta}\big(P(q)\big)F^{\alpha\beta\nu}(q)dqd\tau.$$

Here the point denotes differentiating by the argument and coefficients $\Theta^{\alpha\beta\nu}, Q^{\alpha\beta\nu}, S^{\alpha\beta\nu}, F^{\alpha\beta\nu}$ are represented as

$$\Theta^{\alpha\beta\nu}(x,t,v,y,c) = AJ^3;$$

$$Q^{\alpha\beta\nu}(x,t,v,y,c) = 2A'_tJ^4 + 3AJ'_tJ^3 - BJ^2;$$

$$S^{\alpha\beta\nu}(x,t,v,y,c) = 3A''_{tt}J^5 + 15A'_tJ'_tJ^4$$

$$+11A(J'_t)^2J^3 + 4AJ''_{tt}J^4 - 2B'_tJ^2 - 3BJ'_tJ^2 + CJ;$$

$$F^{\alpha\beta\nu}(x,t,v,y,c) = A''_{tt}J^6 + 9A''_{tt}J'_tJ^5 + 5A'_tJ''_{tt}J^5$$

$$+18A'_t(J'_t)^2J^4 + AJ'''_{ttt}J^5 + 4AJ'_tJ''_{tt}J^4 + 2A(J'_t)^3J^3 - B''_{tt}J^5$$

$$-2B'_tJ_tJ^4 - BJ''_{tt}J^4 - 2B(J'_t)^2J^3 + C'_tJ^2 + C'_tJ'_tJ - D.$$

For brevity we have omitted the indices of A, B, C, D. Namely, everywhere,

$$A = A^{\alpha\beta\nu}(x,t,v,y,c), \qquad B = B^{\alpha\beta\nu}(x,t,v,y,c),$$

$$C = C^{\alpha\beta\nu}(x,t,v,y,c), \qquad D = D^{\alpha\beta\nu}(x,t,v,y,c)$$

and these coefficients are determined as follows:

$$A^{\alpha\beta\nu}(x,t,v,y,c) = \frac{1}{R}\frac{\partial P}{\partial x_\alpha}\frac{\partial P}{\partial x_\beta}\frac{\partial P}{\partial x_\nu};$$

$$B^{\alpha\beta\nu}(x,t,v,y,c) = \frac{1}{R}\left[\frac{\partial^2 P}{\partial x_\alpha \partial x_\beta}\frac{\partial P}{\partial x_\nu} + \frac{\partial^2 P}{\partial x_\alpha \partial x_\nu}\frac{\partial P}{\partial x_\beta} + \frac{\partial^2 P}{\partial x_\beta \partial x_\nu}\frac{\partial P}{\partial x_\alpha}\right]$$

$$-\frac{1}{R_2}\left[\frac{\partial P}{\partial x_\alpha}\frac{\partial P}{\partial x_\beta}\frac{\partial R}{\partial x_\nu} + \frac{\partial P}{\partial x_\beta}\frac{\partial P}{\partial x_\nu}\frac{\partial R}{\partial x_\alpha} + \frac{\partial P}{\partial x_\alpha}\frac{\partial P}{\partial x_\nu}\frac{\partial R}{\partial x_\beta}\right];$$

$$C^{\alpha\beta\nu}(x,t,v,y,c) = \frac{1}{R}\frac{\partial^3 P}{\partial x_\alpha \partial x_\beta \partial x_\nu} - \frac{1}{R_2}\left[\frac{\partial P}{\partial x_\nu}\frac{\partial^2 R}{\partial x_\alpha \partial x_\beta} + \frac{\partial P}{\partial x_\beta}\frac{\partial^2 R}{\partial x_\alpha \partial x_\nu}\right.$$

$$\left. + \frac{\partial P}{\partial x_\alpha}\frac{\partial^2 R}{\partial x_\beta \partial x_\nu} + \frac{\partial R}{\partial x_\nu}\frac{\partial^2 P}{\partial x_\alpha \partial x_\beta} + \frac{\partial R}{\partial x_\beta}\frac{\partial^2 P}{\partial x_\alpha \partial x_\nu} + \frac{\partial R}{\partial x_\alpha}\frac{\partial^2 P}{\partial x_\beta \partial x_\nu}\right]$$

$$+\frac{2}{R^3}\left[\frac{\partial P}{\partial x_\alpha}\frac{\partial R}{\partial x_\beta}\frac{\partial R}{\partial x_\nu}+\frac{\partial P}{\partial x_\beta}\frac{\partial R}{\partial x_\alpha}\frac{\partial R}{\partial x_\nu}+\frac{\partial P}{\partial x_\nu}\frac{\partial R}{\partial x_\beta}\frac{\partial R}{\partial x_\alpha}\right];$$

$$D^{\alpha\beta\nu}(x,t,v,y,c)=-\frac{1}{R^2}\frac{\partial^3 R}{\partial x_\alpha\partial x_\beta\partial x_\nu}+\frac{2}{R_2}\left[\frac{\partial R}{\partial x_\alpha}\frac{\partial^2 R}{\partial x_\beta\partial x_\nu}\right.$$

$$\left.+\frac{\partial R}{\partial x_\beta}\frac{\partial^2 R}{\partial x_\alpha\partial x_\beta}+\frac{\partial R}{\partial x_\nu}\frac{\partial^2 R}{\partial x_\alpha\partial x_\nu}+\frac{6}{R^4}\frac{\partial R}{\partial x_\alpha}\frac{\partial R}{\partial x_\beta}\frac{\partial R}{\partial x_\nu}\right];$$

$$J(x,t,v,y,c)=\left(\frac{\partial P(x,t,v,y,c)}{\partial t}\right)^{-1},$$

where $P(x,t,v,y,c)$ and $R(x,t,v,y,c)$ are determined by (3.21)-(3.23) and the parameter c is either c_p or c_s. Note, that in the case of instantaneous source ($|v|=0$), $J=1$ and the coefficients are as follows

$$\Theta^{\alpha\beta\nu}=A^{\alpha\beta\nu},\quad Q^{\alpha\beta\nu}=-B^{\alpha\beta\nu},\quad S^{\alpha\beta\nu}=C^{\alpha\beta\nu},\quad F^{\alpha\beta\nu}=-D^{\alpha\beta\nu}.$$

The results of the test calculations are given in Figure 3.11.

§3.2 Determining the seismic moment tensor of the seismic centre.

In the previous paragraph we have considered the problems of simulating a seismic vibration field from the sources whose spatial directedness of radiation was described by the seismic moment tensor $M_{ij}(t)$, $i,j=\overline{1,3}$. In this paragraph we shall consider the problems of numerical solution of the inverse problem of determining this tensor and the coordinates of the source hypocentre by the measured volume waves in some points of elastic space.

Setting of the problem. Consider the Lamé system which describes the motion of a homogeneous ideal elastic medium

$$\frac{\partial \sigma_{ij}}{\partial x_j}-\rho\frac{\partial^2 u_i}{\partial t^2}=M_{ij}(t)\frac{\partial}{\partial x_j}\delta(x-y). \tag{2.1}$$

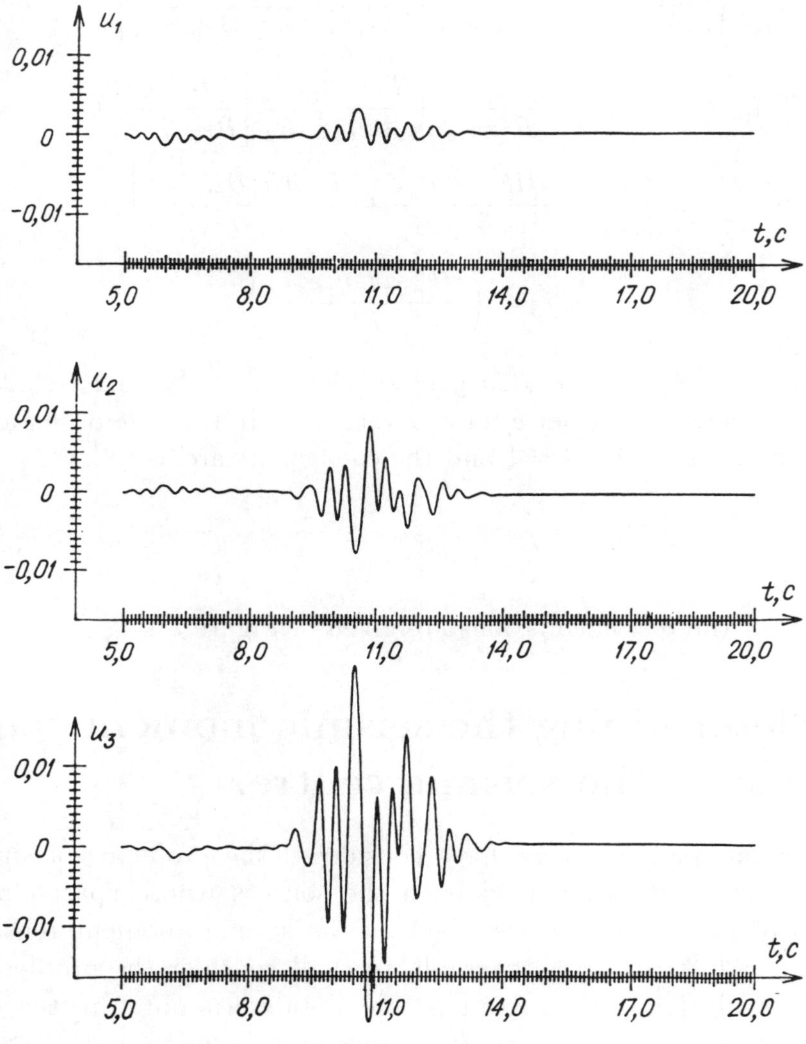

Figure 3.11. Theoretical seismograms for a moving source with seismic moment of the form (3.12). The depth is 10 km ($y = (0, 0, 10)$). The coordinates of the registration point: 30 km, 20 km, 0 km.

Here $i, j = \overline{1,3}$; $x, y \in R^3$; $t \in R^1$; ρ is the medium density; σ_{ij} is the stress tensor which is connected with the shift vector $u(x,t) = (u_1, u_2, u_3)$ as follows

$$\sigma_{ij} = \mu\left(\frac{\partial u_i}{\partial x_j} + \frac{\partial u_j}{\partial x_i}\right) + \lambda\delta_{ij}\frac{\partial u_k}{\partial x_k};$$

μ, λ are the Lamé constants; the repeating indices denote the summing.

The symmetric tensor of rank 2, $M_{ij}(t)$, as was noted in §3.1, is called the seismic moment tensor. It has the dimensions of energy and describes in the terms of equivalent forces the shift gap, the shift destruction of the medium, the transformate change of the volume, the separation crack, and so on (Aki and Richards, 1983). Let t_0 be the start time of the process in the centre; λ, μ, ρ be given constants.

The inverse problem: to determine parameters t_0, y and the tensor $M_{ij}(t)$ from the information

$$v_k(t) = u(x_k, t) + \varepsilon_k(t), \quad k \geq 4. \tag{2.2}$$

Here, $x_k \in R^3$, ε_k is the noise with a normal distribution, a zero average, and the known second moment $G_\varepsilon(x_k, x_{k'})$.

Suppose that there exists a formal procedure $\mathcal{A}$ by means of which we may determine, by the information (3.25), the times of arrival of longitudinal, or transverse waves, or their difference, or altogether

$$\mathcal{A}\Big(v_k(t)\Big) = t_p^k, t_s^k, t_s^k - t_p^k. \tag{2.3}$$

Then the solution to inverse problem (3.24)-(3.26) may found in two stages. First, by information (3.26) to determine the parameters t_0, y. Second, for the known source coordinates and t_0 to determine the dynamic characteristics of the source, i.e., the tensor $M_{ij}(t)$.

In both stages we shall minimize the functional

$$J(\eta) = \Big(w - \mathcal{F}(\eta)\Big)^T V\Big(w - \mathcal{F}(\eta)\Big) + \Omega(\eta) + Sh(\eta). \tag{2.4}$$

Here, $w = (w_1, \ldots, w_N)$ is the measure vector determined by data (3.25); $\eta = (\eta_1, \ldots, \eta_L)$ is the vector consisting of the desired parameters; $\mathcal{F}$ is the vector-solution to equations (3.24) of the same dimension as the information W; $(\,\cdot\,)^T$ denotes the column vector; V is a positive definite symmetric matrix of order $N \times N$, which may coincide with C_ε^{-1}; $\Omega(\eta)$ is the stabilizing Tikhonov functional; $Sh(\eta)$ is the external penalty functional

$$\Theta(\eta) = (\eta - \eta_0)^T R(\eta - \eta_0), \tag{2.5}$$

$$Sh(\eta) = H^-(\eta - \eta_1) + H^+(\eta - \eta_2). \tag{2.6}$$

Here R is a certain $L \times L$ positive definite matrix consisting of regularization coefficients; η_0 is the vector of *a priori* values of the stabilizing functional. Note, that if the correlative matrix of parameter connection $G_\eta = \langle \eta\eta' \rangle$ is given, then as R we may take G_η^{-1}

$$H^-(\eta - \eta_1) = \sum_{m=1}^{L} B_m^- \Theta^-(\eta_m - \eta_{m_1})(\eta_m - \eta_{m_1})^{\nu-}; \tag{2.7}$$

$$H^+(\eta - \eta_1) = \sum_{m=1}^{L} B_m^+ \Theta^+(\eta_m - \eta_{m_1})(\eta_m - \eta_{m_1})^{\nu+}; \tag{2.8}$$

$$\Theta^-(x_m) = \begin{cases} -1, & x_m \le 0, \\ 0, & x_m > 0; \end{cases} \qquad \Theta^+(x_m) = \begin{cases} 1, & x_m \ge 0, \\ 0, & x_m < 0; \end{cases} \tag{2.9}$$

$$B_m^\pm \ge 0, \quad \nu^\pm \ge 2, \quad \eta_{m_2} \ge \eta_{m_1}, \quad m = 1, \ldots, N.$$

Determination of the minimum of (3.27) is achieved by the combination of the Marcwardt method and the regularization quasi-Newton method with step regularization using only first derivatives. Thus, the iterative process of functional minimization is as follows

$$\eta^{k+1} = \eta^k + \alpha_k p^k, \tag{2.10}$$

$$p_k = -A^{-1}(\eta^k, \xi^k) Y(\eta^k, \xi^k). \tag{2.11}$$

Here, α_k is the parameter of step regularization, $1 \leq k \leq k_{\max}$;

$$Y = \left(\frac{\partial \mathcal{F}}{\partial \eta}\right)^T V(\mathcal{F} - w) + R(\eta - \eta_0) + \xi_1 \left(\frac{\partial H^+}{\partial \eta} + \frac{\partial H^-}{\partial \eta}\right); \quad (2.12)$$

$$A = \left(\frac{\partial \mathcal{F}}{\partial \eta}\right)^T V \left(\frac{\partial \mathcal{F}}{\partial \eta}\right) + \Lambda + \xi_1 \left(\frac{\partial^2 H^+}{\partial \eta^2} + \frac{\partial^2 H^-}{\partial \eta^2}\right). \quad (2.13)$$

The vector ξ in (3.33), (3.34) is the vector of all free parameters of functional (3.27) and iterative process (3.36) identified on each step by k

$$\xi^{k+1} = \xi^k + \Delta\xi\left(\eta^1, \ldots, \eta^k\right). \quad (2.14)$$

Thus, when solving inverse problem (3.24), (3.25), we use the adaptive approach, when the parameters of the penalty functional of the Tikhonov stabilizor are formed on each iterative step. The combination of methods is achieved by introducing the independent matrices Λ and R. Really, in the case when $\Lambda = R$, we obtain the regularized quasi-Newton method. When $R = 0$, the algorithm (3.35), (3.36) is the generalized Marcwardt method (Λ is arbitrary in contrast to the standard method, when $\Lambda = \lambda I$).

The parameter α_k in (3.35) in each step of the iterative process is chosen maximal from the set $\{\alpha_k |\, 0 < \alpha_k < \alpha_0\}$ such that

$$J\left(\eta^k + \alpha_k p^k\right) - J\left(\eta^k\right) \leq \varepsilon \alpha_k \left(Y, p^k\right), \quad (2.15)$$

where $0 < \varepsilon \leq 1/2$. For $\alpha_0 = 1$ the method with the step regularization coincides with the classical one (Pshenichny and Danilin, 1975). If we have the case $\alpha_0 > 1$, first, we check the global minimum in the antigradient direction. In there is no global minimum, then we use the standard method of α subdivision.

Iterative process (3.35), (3.36) finishes if either $k > k_{\max}$, or if differences between the values of certain parameters become by absolute value less than certain threshold numbers δ_i, $i = 1, \ldots, L$.

The choice of optimal values of regularization parameters. One of the most complicated issues, when solving inverse problems, is the solution regularization. This means the concordance of the regularization parameter matrix G with the given noise level in the data v_k. In the considered problem we use the residual method (Tikhonov *et al.*, 1983) of choosing the regularization parameter. We shall illustrate this method taking as the example the test problem. Let $G = \alpha I$, where I is the unit matrix $n \times n$. On the basis of solving equation (3.24) (Erokhin and Bortnikov, 1987, 1988) there were modeled the data (3.25) for four points of observation with the coordinates $x_{k3} = 0$, $x_{k2} = 0$, $x_{11} = 1$ km, $x_{21} = 5$ km, $x_{31} = 8$ km, $x_{41} = 13$ km, $k = 1, \ldots, 4$.

The noise was modeled by random numbers with a normal distribution and zero mean value. The test calculations were carried out for the cases when the relative noise level ε in the data, taken as $\varepsilon = ||v_\varepsilon - u||/||u||$, where $||\varepsilon||$ is the mean square norm, was 0, 10, 20 and 40%. The source parameters were $y_1 = y_2 = y_3 = 0$, $t_0 = 0$, $v = 0$. The tensor was taken arbitrarily as

$$M_{ij}(t) = \delta(t) \begin{pmatrix} 1 & 4 & 5 \\ 4 & 2 & 6 \\ 5 & 6 & 3 \end{pmatrix}.$$

The concordance procedure for regularization parameter α in this case is determining the optimal parameter $\alpha = \alpha_{\text{opt}}$, where for the given noise level ε the relative error for determining parameters δ ($\delta = ||\eta_\varepsilon - \eta||/||\eta||$) is minimal.

In Figure 3.12 for the freguencies 1 and 20 Hz are given the dependences δ on α for different noise levels in the case when the vector η consists of the components of $\hat{M}_{ij}(\omega)$. It was assumed that the centre coordinates and t_0 are explicitly determined in the first stage. Solution to the problem in the second stage was obtained in the spectral domain, i.e., the value of $\hat{M}_{ij}(\omega)$ was determined for each of 256 frequencies. The analysis of test calculations have shown that α_{opt} depends on mutual disposition of the seismic centre and observation stations from the noise value, and on the frequency ω. For example, it was noted that when the frequency or the noise grows, then α_{opt} also grows and may attain 10^{-1}.

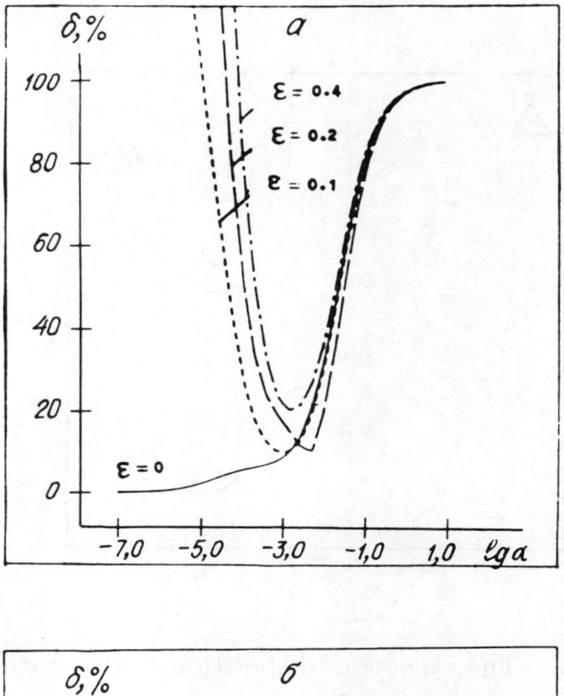

Figure 3.12. Dependence of relative errors of determining the components of the seismic moment tensor on the regularization parameter for different noise levels in data, $R = 5$km

a) $\omega = 20$ Hz; *b*) $\omega = 1$ Hz.

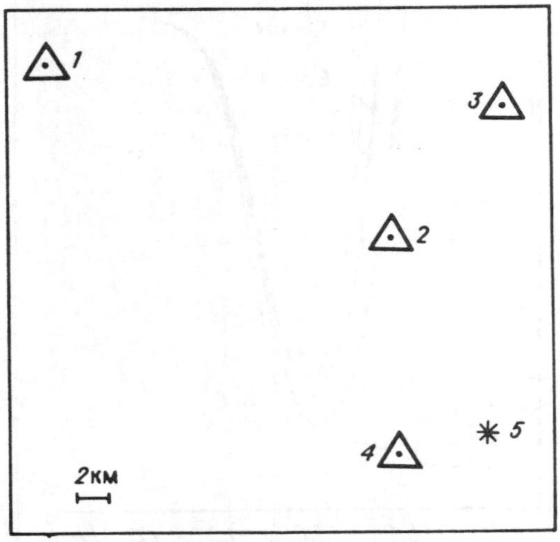

Figure 3.13. The scheme of disposition of the station 1-4 and the earthquake centre -5. The polygon sizes $20' \times 30'$.

It is interesting that if we have a small noise in data ($\varepsilon \leq 10\%$), we cannot obtain a stable solution without regularization (i.e., for $\alpha = 0$). This is shown in Figure 3.12.

Denote, further, by u_α the solution to (3.24) with the parameters y, t_0 and the tensor $M_{ij}(t)$ obtained by minimizing functional (3.27) for a certain fixed regularization parameter α (recall, that $G = \alpha I$). Then, denote by $\delta_\alpha^\varepsilon$ the relative value of residual u_α and the experimental data v_ε for the given noise level $\delta_\alpha^\varepsilon = ||u_\alpha - v_\varepsilon||/||v_\varepsilon||$.

The results of approving the method on the experimental record. Determining the shift, separation, and hydrostatic stresses in the centre. The suggested method for numerical solution of the inverse problem was tested by processing seismic earthquake records of the earthquake of the class $K = 7$ which occured in Altay in 1985. It was registered by four three-component stations "Region" (Figure 3.13).

The coordinates of the centre were determined on the basis of information on the difference of incoming longitudinal and transverse waves. The centre depth was $y_3 = 14.4 \pm 0.5$ km. The medium parameters had the following values $\rho = 2.7$ g/cm^3, $c_p = 6.1$ km/s, $c_s = 3.5$ km/s. In Figure 3.14 we may observe one of the components of numerical record of the station number 4 and its spectrum. The time is taken with $t_0 = 0$.

On the basis of the methods developed for the given position of the stations, the optimal regularization parameters were calculated for 256 frequencies assuming that the noise level in data is less than 30%. From the results of test calculations, the accuracy of determining source parameters in this case will be higher than 30%.

In Figure 3.15 are given the spectral densities of the seismic moment tensor $\hat{M}_{ij}(\omega)$. As we may see, in the centre parallel with the shift stresses there are hydrostatic stresses. Their separation is possible by reduction in each moment of the stresses to the principal axes in the centre. These stresses in their turn are determined by eigenvectors $p(t) = (p_1(t), p_2(t), p_3(t))$ of the tensor $M_{ij}(t)$

$$M_{ij}(t) \to M'_{ij}(t) = \begin{pmatrix} a_1(t) & 0 & 0 \\ 0 & a_2(t) & 0 \\ 0 & 0 & a_3(t) \end{pmatrix},$$

$$a_1(t) > a_2(t) > a_3(t).$$

Further interpretation is possible on the basis of separation of the tensor $M'_{ij}(t)$, the ball tensorm, and the stress deviator

$$M'_{ij}(t) = \frac{1}{3}\begin{pmatrix} p_0 & 0 & 0 \\ 0 & p_0 & 0 \\ 0 & 0 & p_0 \end{pmatrix} + \begin{pmatrix} a_1 - \frac{1}{3}p_0 & 0 & 0 \\ 0 & a_2 - \frac{1}{3}p_0 & 0 \\ 0 & 0 & a_3 - \frac{1}{3}p_0 \end{pmatrix},$$

$$p_0(t) = \sum_{k=1}^{3} a_k(t).$$

Hence, on the basis of the extremal problem for the tangent stress $\sigma(t)$, we may determine the normal vector $n(t)$ to the platforms of maximal tangent stresses, the shift stresses $\sigma^{\tau}_{\max}(t)$ in them, and the separation stresses $\sigma^n(t)$ which are determined as follows

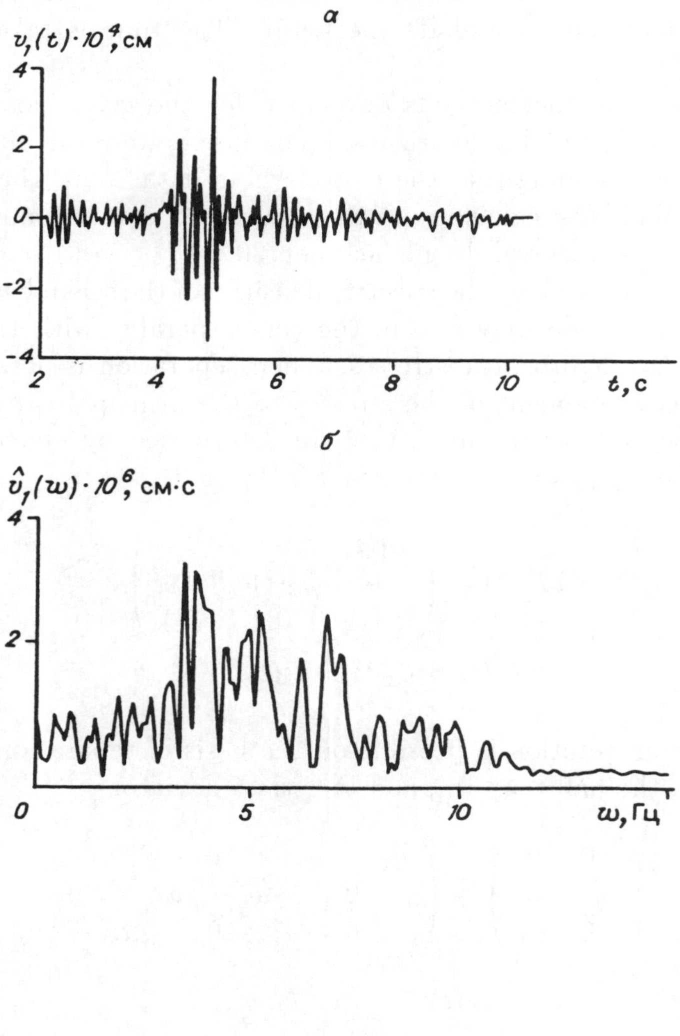

Figure 3.14. The record of one of the components of the shift vector in the station number 4 (a) and its amplitude spectrum (b).

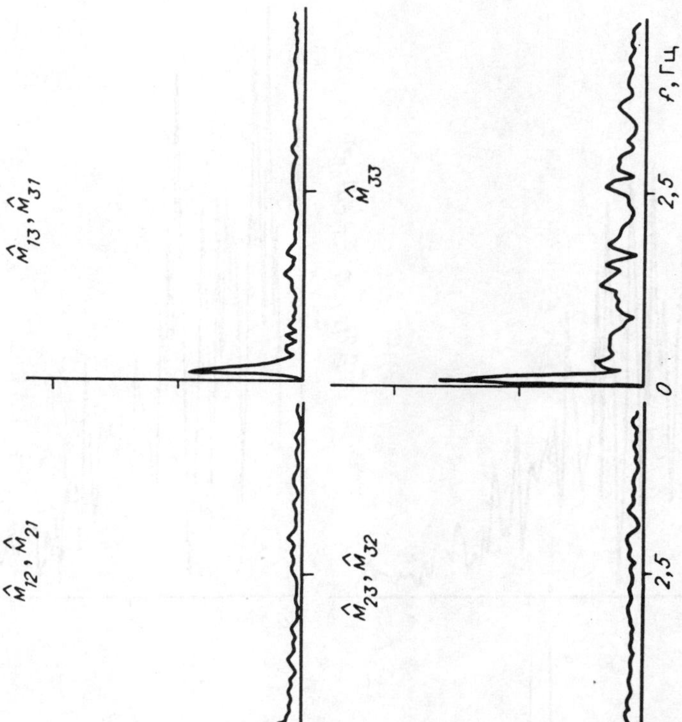

Figure 3.15. Spectral densities of components of the obtained seismic moment tensor.

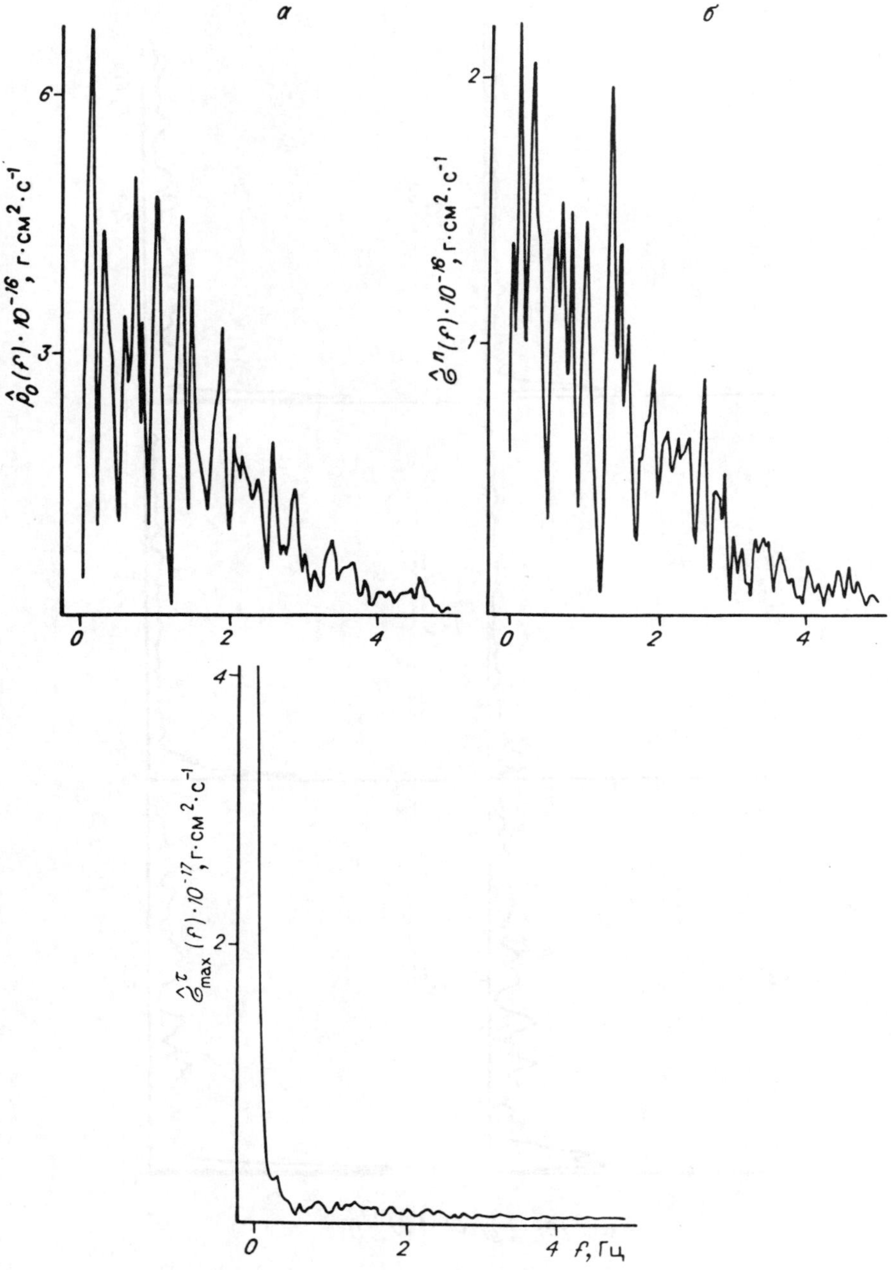

Figure 3.16. Spectral densities of the hydrostatic tensor p_0, of separation stresses σ^n, and shift streses $\sigma^\tau_{\max}$.

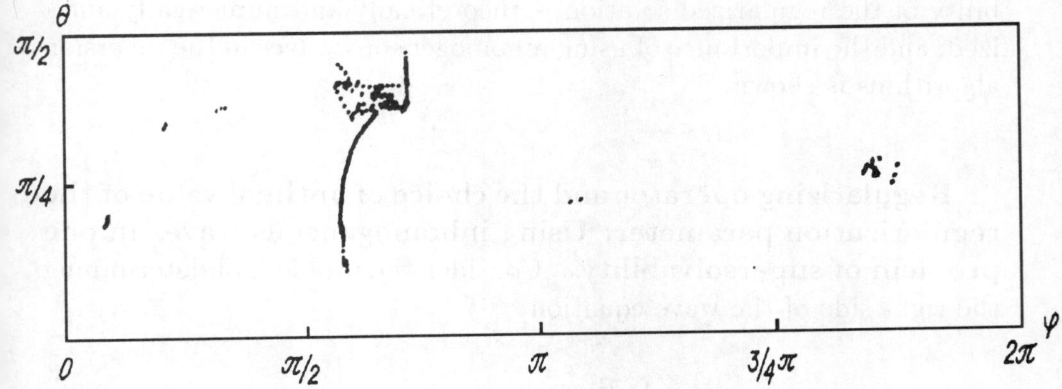

Figure 3.17. Projections of the normal vector to platforms of tangent stresses of the centre in the unit hemisphere after $\Delta t = 0.05$ s.

$$\sigma^{\tau}_{\max}(t) = \frac{a_1(t) - a_3(t)}{2},$$
$$\sigma^{n}(t) = \frac{a_1(t) + a_3(t)}{2}.$$

From Figure 3.16 we see that in the centre the shift stresses predominate. The domains of concentration of normal vectors to platforms of maximal tangent stresses show the orientation in the space of the platforms, where, possibly, had been a gap in the centre.

§3.3 The integral transformations method in inverse problems of source determination.

In the previous paragraph we have considered the theoretical problems of applying the integral transformations method in the problems of determining right sides of differential equations. Constructive algorithms of inversion for certain right sides were suggested. In this section we consider the numerical realization of the suggested algorithms; the stability of the regularized solution is theoretically and numerically analized; and the importance of using inhomogeneous waves in the inversion algorithms is shown.

Regularizing operator and the choice of optimal value of the regularization parameter. Using inhomogeneous waves in one problem of supersolvability. Consider the problem of determining the right side of the wave equation

$$\Delta u = \frac{1}{c^2}\frac{\partial^2 u}{\partial t^2} = \delta^{(k)}(z-h)f(x,t), \tag{3.1}$$

where $k = 0, 1$; $x \in R^2$; $z, t \in R^1$, by the information given in the plane $z = 0$

$$u(x,z,t)\Big|_{z=0} = u_0(x,t). \tag{3.2}$$

According to results of Erokhin (1982), the solution to such a problem is given by the formula

$$\hat{f}(\lambda,\omega) = \hat{u}_0(\lambda,\omega)e^{|h|\sqrt{s(\lambda,\omega)}}s^{\frac{1-k}{2}}(\lambda,\omega), \tag{3.3}$$

where λ, ω are the spectral variables of the Fourier transformation corresponding to the variables x, t ($\lambda \in R^2$, $\omega \in R^1$);

$$s(\lambda,\omega) = \lambda_1^2 + \lambda_2^2 - \left(\frac{\omega}{c} + i\alpha\right)^2, \quad \mathrm{Re}(\sqrt{s}) > 0.$$

The absorption coefficient $\alpha (0 < \alpha < 1)$ is introduced here only for separation of the divergent waves, i.e., for correct determination of the sign of the imaginary part $\sqrt{s}$, which is found in this case from the relation

$$\mathrm{Im}(\sqrt{s})\mathrm{Re}(\sqrt{s}) = -\alpha\frac{\omega}{c}.$$

As, by the definition, $\mathrm{Re}(\sqrt{s}) > 0$, then, as may be seen from (3.41), the solution to such a problem is classically unstable, since the inversion operator is unbounded. Therefore, we introduce the notion of the regularizing operator $\hat{B}_\Lambda$, in the sense of Ivanov *et al.* (1978) and Tikhonov and Arsenin (1979), which is the approximation of the initial unbounded operator $\hat{B}$.

$$\hat{B}_\Lambda(\lambda,\omega) = \begin{cases} s^{\frac{1-k}{2}}(\lambda,\omega)\exp(|h|\sqrt{s(\lambda,\omega)}), & \lambda \le \frac{\omega^2}{c^2} + \Lambda^2, \\ 0, & \lambda > \frac{\omega^2}{c^2} + \Lambda^2 \end{cases} . \quad (3.4)$$

Here $\lambda^2 = \lambda_1^2 + \lambda_2^2$; $|\omega| \le \Omega$; $\Lambda \ge 0$.

In Figure 3.18 in the plane ω, λ the continuous line bounds the domain where this operator is not zero. The horizontal prime line denotes the domain where for $k = 1$: $\alpha = 0$ and $\left\|\hat{B}_\Lambda\right\| = 1$.

Thus, for $\Lambda = 0$ in the recovery operator we use only homogeneous waves (Klayerboat, 1981) (strong regularization).

It may be shown that the action of $\hat{B}_\Lambda$ for $\Lambda > 0$ is represented as the sum of two convolutions

$$f(r,\omega) = u_0(r,\omega) * K_0(r,\omega) + u_0(r,\omega) * K_\Lambda(r,\omega).$$

Here,

$$K_\Lambda(r,\omega) = \int_0^\Lambda e^{|h|\tau} J_0\left(r\sqrt{\tau^2 + \frac{\omega^2}{c^2}}\right)\tau d\tau - \int_0^\infty e^{i|h|\tau} J_0\left(r\sqrt{\tau^2 + \frac{\omega^2}{c^2}}\right)\tau d\tau;$$

$$K_0(r,\omega) = \frac{d}{dh}G^*(r,\omega),$$

Here, $r^2 = x_1^2 + x_2^2 + h^2$; $G^*(r,\omega)$ is the Green function for the Helmholtz equation.

The first convolution corresponds to using only homogeneous waves in the operator, and the second, to using inhomogeneous waves also in the algorithm.

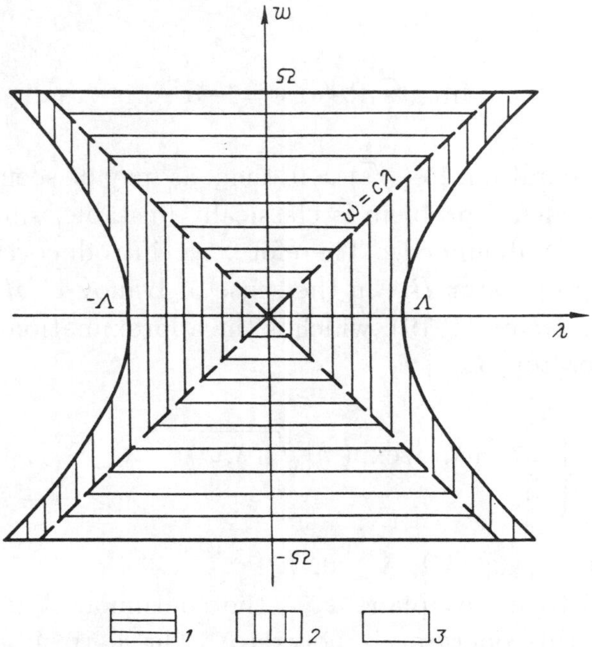

Figure 3.18. The domain of determination of the regularizing operator $\hat{B}_\Lambda$.

Domains: 1 - homogeneous waves ($\omega \le c|\lambda|$); 2 - inhomogeneous waves ($c|\lambda| < |\omega| < c\sqrt{\Lambda^2 - \lambda^2}$); 3 - zero value of the operator.

Now we investigate the stability of determining the right side of the form (3.40) by means of the regularized operator $\hat{B}_\Lambda$ (3.42) in the case when the function $f(x,t)$ is a regular function from $L_2(R^3)$.

First, let $k = 1$ and suppose the problem is considered for a certain fixed frequency ω_0. Denote by the well-posedness set M the set of such functions $f(x, \omega_0) \in L_2(R^2)$ that $\left|\left|f\right|\right| < c_0$. Then, as follows from (3.41), the function $u_0(x, \omega)$ also belongs to $L_2(R^2)$.

Further, let $0 \le h \le H$, and instead of an explicit field, suppose we know the approximate $u_0^\delta(x, \omega_0) \in N$, where $N = AM$ is the image of the set M (A is the solution to the problem), such that

$$\left\|u_0^\delta - u_0\right\| \leq \delta.$$

Also, suppose we know only an approximate value of h : h_ε, such that $\|h_\varepsilon - h\| \leq \varepsilon$. Denote by $\hat{B}_\Lambda^\varepsilon$ the operator (3.42) for $h = h_\varepsilon$. It is required to estimate the accuracy of determining the right side when the regularizing operator B_Λ^ε acts on the information u_0^δ.

Introduce the quantitative characteristic of accuracy of the method in question

$$\left\|B_\Lambda^\varepsilon u_0^\delta - f\right\| \leq \left\|\left(\hat{B}_\Lambda^\varepsilon - B_\Lambda\right)\hat{u}_0^\delta\right\| + \left\|\hat{B}_\Lambda \hat{u}_0 - \hat{f}\right\| + \left\|\hat{B}_\Lambda\left(\hat{u}_0^\delta - \hat{u}_0\right)\right\|. \quad (3.5)$$

Using the Parseval equality (Danford and Schwarz, 1962), we obtain the estimate

$$\Delta_\Lambda = \sup\left\{\left\|B_\Lambda^\varepsilon u_0^\delta - f\right\| \,\middle|\, \left\|u_0^\delta - u_0\right\| \leq \delta,\ \left|h_\varepsilon - h\right| \leq \varepsilon,\ u_0^\delta \in N\right\}. \quad (3.6)$$

The first term in the right side of (3.43) characterizes the error induced by the inaccurate value of h for operator (3.42); the second term characterizes the error of replacing B by the bounded B_Λ. Finally, the third term gives the error induced by the noise in the initial data u_0.

On the basis of such estimate we may show that the maximal error is as follows:

$$\Delta_\Lambda = e^{H\Lambda}\left(e^{\varepsilon\Lambda} - 1\right)c_0 + c_0 - \left\|\hat{f}\right\|_\Lambda + \delta e^{H\Lambda}, \quad (3.7)$$

where

$$\left\|\hat{f}\right\|_\Lambda^2 = \int_{\lambda^2 \leq \omega_0^2/c^2 + \Lambda^2} \hat{f}^2(\lambda; \omega_0)\, d\lambda.$$

Formula (3.45) shows that for $\Lambda > \Lambda^*$, which is obtained from the inequality

$$e^{\varepsilon\Lambda}c_0 - c_0 \geq \delta$$

and is as follows

$$\Lambda^* = \frac{\ln\left(1 + \frac{\delta}{c_0}\right)}{\varepsilon},$$

the error in determining the source caused by the inaccurate value of h will exceed the error caused by inaccuracy of the field $u_0(x, \omega_0)$ representation.

On the basis of (3.45) we may find such Λ_{opt} that the corresponding method will be optimal

$$\Delta_{\text{opt}} = \inf_{\Lambda \geq 0} \{\Delta_\Lambda\}.$$

Evidently, determining Λ_{opt} is possible only for *a priori* knowledge of $\|\hat{f}\|_\Lambda$, or its behavior in infinity. So, for example, if for the given wave number $k_0 = \omega_0/c$ we know that $\|\hat{f}\|_\Lambda = \|\hat{f}\|$, then, according to (3.45), we obtain

$$\Delta_{\text{opt}} = \inf_{\Delta \geq 0} \left\{ e^{H\Lambda} \left(e^{\varepsilon\Lambda} - 1 \right) c_0 + \delta e^{H\Lambda} \right\} = \delta. \tag{3.8}$$

In this case, $\Lambda_{\text{opt}} = 0$. In practice, the sources have a compact support with respect to x_1, x_2; therefore, their spectrum is not finite. Nevertheless, qualitatively, we may establish the following regularity: the more narrow is the spatial spectrum of the source, the more stable is our solution for the inverse problem, if we use in the algorithm only homogeneous waves ($\Lambda = 0$).

Returning to (3.46), we note that this estimate includes the following three particular cases

a) $B_\Lambda^\varepsilon = B_\Lambda$, $\delta \neq 0$, then

$$\left\| B_\Lambda u_0^\delta - f \right\| \leq \delta e^{H\Lambda} + c_0 - \|f\|_\Lambda;$$

b) $B_\Lambda^\varepsilon = B_\Lambda$, $\delta = 0$, then

$$\left\|B_{\Lambda} u_0 - f\right\| \le c_0 - \left\|f\right\|_{\Lambda};$$

c) $B_{\Lambda}^{\varepsilon} \neq B_{\Lambda}$, $\delta = 0$, then

$$\left\|B_{\Lambda}^{\varepsilon} u_0 - f\right\| \le e^{H\Lambda}\left(e^{\varepsilon\Lambda} - 1\right) + c_0 - \left\|\hat{f}\right\|_{\Lambda}.$$

The first estimate for a spectral case of the f space for the problem of the field extension was obtained by Marchuk (1978). In this work, the author introduced the notion of a conditionally optimal method to which the parameter $\tilde{\Lambda}_{\text{opt}}$ corresponds. The value of this parameter is determined from the condition of stability of recovery error stipulated by the inaccuracy

$$\sigma = \left\|\hat{B}\Lambda\right\|\delta = e^{H\Lambda}\delta.$$

Hence follows

$$\tilde{\Lambda}_{\text{opt}} = \left[\ln(\sigma/\delta)/H\right]. \qquad (3.9)$$

This notion may be extended to the more general case considered above - estimate (3.46). So, substituting (3.46) into (3.47), we obtain

$$\tilde{\Delta}_{\text{opt}} = \frac{\sigma}{\delta}\left[\left(\frac{\sigma}{\delta}\right)^{\varepsilon/H} - 1\right] c_0 + c_0 - \left\|\hat{f}\right\|_{\tilde{\Lambda}_{\text{opt}}} + \sigma.$$

Hence, for the functions satisfying the inequality

$$c_0 - \left\|\hat{f}\right\|_{\Lambda} \le \frac{1}{\sqrt{k_0^2 + \Lambda^2}},$$

we have

$$\tilde{\Delta}_{\text{opt}} = \frac{\sigma}{\delta}\left[\left(\frac{\sigma}{\delta}\right)^{\varepsilon/H} - 1\right] c_0 + \left(k_0^2 + \frac{\ln(\sigma/\delta)}{H^2}\right)^{-1/2} + \sigma. \qquad (3.10)$$

From (3.48), in the particular case $\varepsilon = 0$, the conditional stability estimate follows (Marchuk, 1978).

We should consider how the regularizing operator (3.42) acts for $k = 1$ for the frequencies $-\Omega \le \omega \le \Omega$.

The correctness set in this case we define as the set of $f(x,t) \in L_2(R^3)$ such that

$$\left|\left|f\right|\right|_{L_2(R^3)} \leq c_0.$$

Further, we assume that $0 \leq h \leq H$ and $\left|\left|u_0^\delta - u_0\right|\right| \leq \delta$. Then, introducing the norm

$$\left|\left|\hat{f}\right|\right|_\lambda^z = \int_{-\Omega}^{\Omega} \int_{\lambda^2 \leq \frac{\omega^2}{c^2} + \Lambda^2} \hat{f}(\lambda, \omega) d\lambda d\omega,$$

we may show that (3.44) holds, and Δ_Λ is represented as (3.45).

Depending on the behavior of $\hat{f}$, we may also find the optimal value of the regularizing parameter Λ_{opt} which corresponds to the minimal error of recovery; and, also, the conditionally optimal value $\tilde{\Lambda}_{\text{opt}}$ for which the estimate of recovery accuracy analogous to (3.48) may be obtained.

Now, let $k = 0$. The regularizing operator in this case, as before, is calculated by formula (3.42). Considering for this operator the analogous problem of solution stability we may show that Δ_Λ (see (3.44)) for the fixed frequency has the form

$$\Delta_\Lambda = e^{H\Lambda}(e^{\varepsilon\Lambda} - 1)c_0 + c_0 - \left|\left|\hat{f}\right|\right|_\Lambda + \Lambda e^{H\Lambda} \cdot \delta. \qquad (3.11)$$

This expression differs from (3.45) for $k = 1$ by an additional factor Λ in the last term. This leads to the fact that for $\Lambda = 0$, the accuracy of the method (in the L_2 norm) does not depend on the data error and the explicit value h, since in this case $\Delta_\Lambda = c_0 - \left|\left|\hat{f}\right|\right|_\Lambda$; and the accuracy depends only on how compact is the function f support. So, if $\left|\left|\hat{f}\right|\right|_{\Lambda=0} = \left|\left|\hat{f}\right|\right|$, then, analogous to (3.46), in our case we have

$$\Delta_{\text{opt}} = \inf_{\Lambda > 0} \left\{ e^{H\Lambda}(e^{\varepsilon\Lambda} - 1)c_0 + \Lambda e^{H\Lambda}\delta \right\} = 0,$$

where, also, $\Lambda_{\text{opt}} = 0$.

When determining the conditionally optimal method, we may use the methods stated above. It should be noticed that $\tilde{\Lambda}_{\text{opt}}$ in this case has no analytic representations analogous to (3.47), since it is the solution to the traditional equation

$$\ln \frac{\sigma}{\delta} + \ln \tilde{\Lambda}_{\text{opt}} = H\tilde{\Lambda}_{\text{opt}}.$$

As for the stability of the recovery problem in the frequency interval $-\Omega \leq \omega \leq \Omega$, it should be noticed that in this case (3.49) also holds. It is only necessary to replace $\left\|h\right\|_{\Lambda}$ by $\left\|\hat{h}\right\|_{\lambda}$. All the considerations are analogous.

Remark. For other sets $Q \ni z$, $T \ni z$ (Erokhin, 1982) and $k \geq z$ we also may construct the regularizing operator (3.42) and obtain stability estimates similar to the above estimates.

Now we consider some problems of numerical realization of algorithm (3.41), (3.42). For simplicity we shall investigate first the case of the fixed frequency $\omega_0 = 2\pi c/k_0$.

It is known that passing from a continuous argument to a discrete one corresponds to the situation when instead of a finite spectrum we obtain a periodic spectrum with period equal to the inverse digitization value (Rabiner and Gould, 1978; Tikhonov and Arsenin, 1979). Thus, for digitization in the spatial domain x_1, x_2 with the step $\Delta x_1 = \Delta x_2 - \Delta z$, according to the Kotel'nikov theorem (Rabiner and Gould, 1978), we assume *a priori* that the spectrum of the function for digitization is bounded by the value $|\Theta| = \frac{\pi}{\Delta z}$.

In order that the information $u_0(x, \omega_0)$ does not change until we apply the regularizing operator B_Λ it is necessary that $|\Theta| \geq \sqrt{\Lambda^2 + k_0^2}$, or, in the spatial domain

$$\Delta z \leq (\Lambda^2 + k_0^2)^{-\frac{1}{2}} \leq \lambda_0/2,$$

i.e., the digitization step must be less than one half wavelength of radiation. In the spectral domain, the digitization step

$$\Delta\lambda_1 = \Delta\lambda_2 = \frac{2\Theta}{N-1} = \frac{2\pi}{\Delta z(N-1)},$$

where N is number of references in the line of receiving aperture, which we everywhere below we will take as the square with side $2a$.

Consider the regularizing operator B_Λ for $\Lambda = 0$. In this case it is similar to the operator of wave field extension which is widely met in seismic exploration (Timoshin, 1978; Tsibulchik, 1975). Suppose, first, that the support of $\hat{f}(\lambda, \omega_0)$ lies in the disk of radius k_0. Then, according to (3.45), the error in determining $f(x, \omega_0)$ is dependent only on the inaccuracy of the initial information u_0^δ

$$\left\| u_0^\delta - u_0 \right\| \le \delta.$$

This inaccuracy arises because of imperfections in the receiving aperture and the noise.

Thus, if the noise is absent, we may achieve any accuracy of determination of the source by increasing the number of references N for the fixed digitization step $\Delta z \le \lambda_0/z$. But, as a rule, the support of $f(x, \omega_0)$ by the variables x_1, x_2 is finite, and, thus, the spectrum of $\hat{f}(\lambda, \omega_0)$ is *a priori* wider than the disk of radius k_0. Then, according to (3.45), even if the data are given exactly, we cannot by means of the operator $\hat{B}_0 = \hat{B}_{\Lambda=0}$ decrease the error stipulated by eliminating the initial information. The latter, according as (3.45), has the estimate

$$\left\| B_0 u_0^\delta - f \right\| \le \left\| f \right\|_{L_2} - \left\| \hat{f} \right\|_{\lambda^2} \le k_0^2.$$

So, we consider the action of B_Λ for $\Lambda > 0$, i.e., using in the recovery algorithm the inhomogeneous waves. In this case, as was shown before, there must exist a certain optimal value of the regularization parameter $\Lambda_{\text{opt}} > 0$, for which the error of determining the right side will be minimal.

Dependencies of relative error of determining the function f

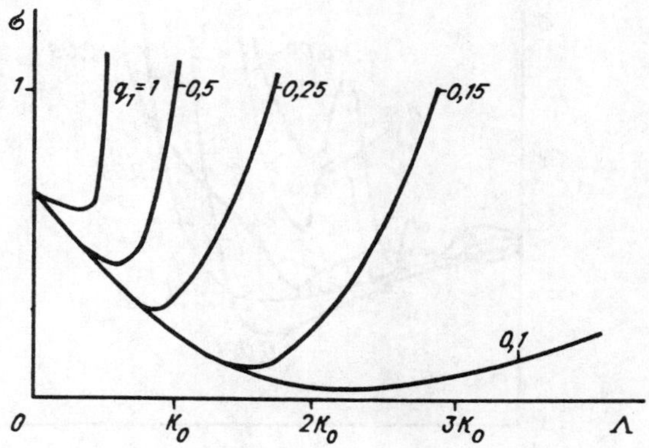

Figure 3.19. Dependence of relative error of recovery from the regularization parameter for disfocusing.

$$\sigma = \frac{\left\| f - B_0 u_0^\delta \right\|}{\left\| f \right\|}$$

on the regularization parameter $\Lambda \geq 0$ for different values of relative error

$$q_1 = \frac{|h_\varepsilon - h|}{h} = \frac{\varepsilon}{h}$$

are represented in Figure 3.19.

The dependencies for relative error in initial information

$$q_2 = \frac{\left\| u_0^\delta - u_0 \right\|}{\left\| u_0 \right\|} = \frac{\delta}{\left\| u_0 \right\|}$$

are given in Figure 3.20. Here $h = \lambda_0$,

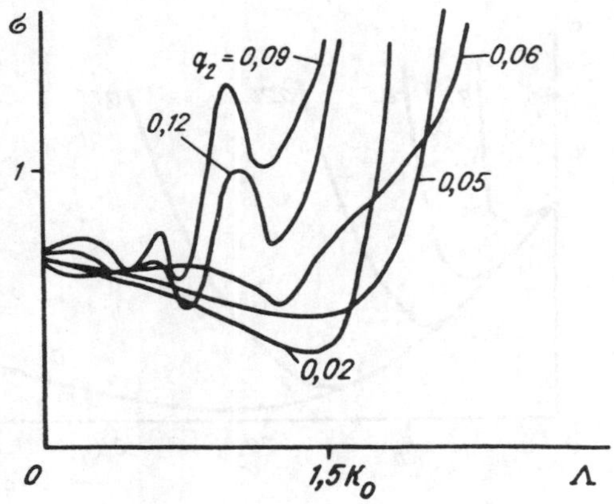

Figure 3.20. Dependencies on the regularization parameter of relative error of recovery in the presence of the noise for initial information.

$$f(x, \omega_0) = \exp\left\{-\frac{\omega_0}{c}\beta^2\left(x_1^2 + x_2^2\right)\right\}; \quad \beta = 4, \quad N = 64.$$

We may see from the Figures that for each q_1 and q_2 there corresponds its own value of the regularization parameter Λ_{opt} for which the mean square error of the source determination is minimal. Moreover, according to theoretical estimate (3.45), if the relative estimates q_1 and q_2 increase, the value of the optimal regularization parameter decreases, tending to zero. Thus, for large errors in initial data, the optimal inversion operator is the "strongly regularized" operator of wave field extension used in seismics (Timoshin, 1978; Tsibulchik, 1975b).

As an additional illustration of the regularizing properties of operator (3.42) we may see in Figure 3.21 the images of two recovered point sources for different values of the regularization parameter $\Lambda \geq 0$. The sources are disposed in the plane $z = h$. Here, also, $a = 4\lambda_0$, $h = \lambda_0$. As we may see from this Figure, for $\Lambda = 1.5K_0$, the difference between

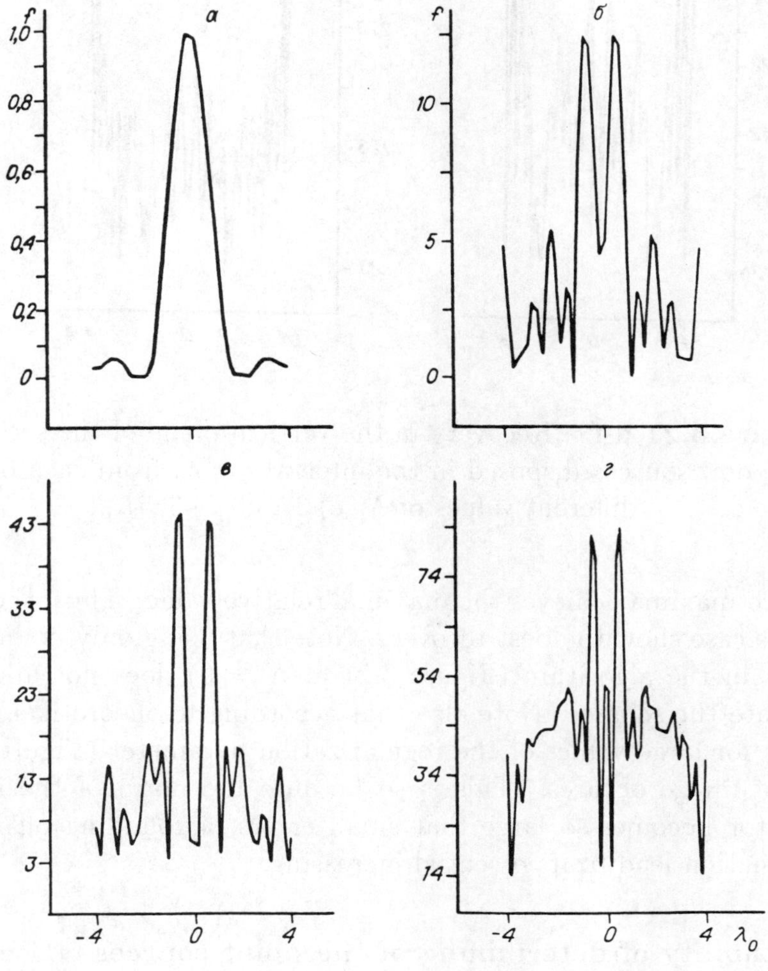

Figure 3.21 a-d. Recovery in the vertical plane of the sections of two point sources disposed in the interval $l = \lambda_0$ from each other for different values of Λ.

a) 0; *b*) $0,75K_0$; *c*) $1,5K_0$; *d*) $2,3K_0$.

The distance to receiving aperture $h = \lambda_0$; the side of aperture $8\lambda_0$.

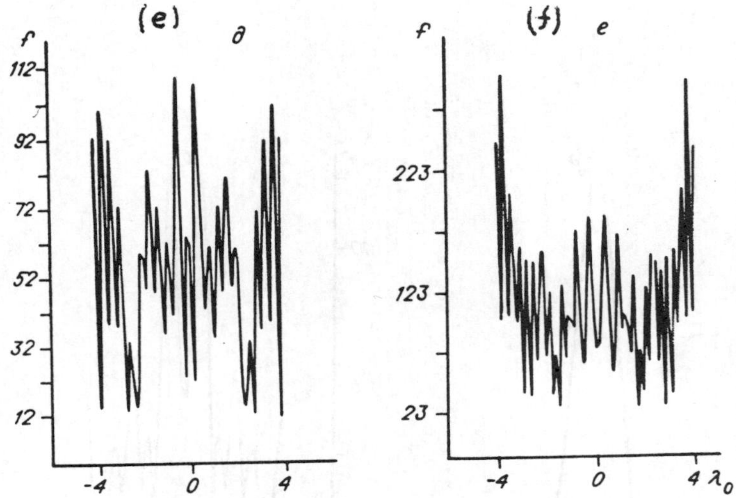

Figure 3.21 e-f. Recovery in the vertical plane of the sections of two point sources disposed in the interval $l = \lambda_0$ from each other for different values of Λ: $e)$ $3K_0$; $f)$ $3,5K_0$.

two maxima achieves the maximal relative value. Thus, the sources in this case show the best recovery. Note that using only homogeneous waves in the algorithm (Figure 3.21 a, $\Lambda = 0$) does not allow us to separate the sources. Note also that according to theoretical eatimate (3.44) for large values of the regularization parameter (Figure 3.21, e) loss of image occures. This is so because the norm of the inversion operator becomes so large that small errors of rounding off in initial information lead to its essential increasing.

Stability of determining of the point sources. Consider for a fixed frequency ω_0 the action of operator (3.42) for $k = 1$ and $\Lambda = 0$ on the function $u_0^r(x_1, x_2, \omega_0)$ such that

$$u_0^r(x_1, x_2, \omega_0) = u_0(x_1, x_2, \omega_0)\mathrm{rect}\left(\frac{x_1}{2a}\right)\mathrm{rect}\left(\frac{x_2}{2a}\right).$$

Suppose we know the approximate value h_ε, and the sources are as follows

$$f(x_1, x_2, \omega_0) = \sum_{k=1}^{N} \varphi_k(\omega_0)\delta\left(x_1 - y_1^k\right)\delta\left(x_2 - y_2^k\right). \qquad (3.12)$$

According to formula (3.42), the action of $B^{\varepsilon}_{\Lambda=0} = B_0^{\varepsilon}$ in the spatial domain will have the convolution form

$$f^r(x, \omega_0) = \int_{R^2} B_0^{\varepsilon}\left(x - x', \omega_0\right) u_0^r\left(x', \omega_0\right) dx', \qquad (3.13)$$

where B_0^{ε} is as follows

$$B_0^{\varepsilon} = \frac{d}{dh_\varepsilon}\frac{e^{-ik_0 R}}{R},$$

and where , in its turn, $k_0 = \omega_0/c$, $R^2 = (x - x')^2 + h_\varepsilon^2$.

Expanding the function R into the Taylor series

$$R = h_\varepsilon + \frac{(x - x')^2}{2h_\varepsilon} + \cdots = h_\varepsilon + \frac{x'^2}{2h_\varepsilon} + \frac{x^2}{2h_\varepsilon} - \frac{xx'}{h_\varepsilon} + \cdots$$

and taking only the first two terms of the series, we obtain to within non-essential multipliers the approximate B_0^{ε}

$$B_0^{\varepsilon} \approx \exp\left\{ik_0(x_1'x_1 + x_2'x_2) + \frac{(x_1')^2 + (x_2')^2}{2h_\varepsilon}\right\}.$$

Noting the fact that the right side of the wave equation has special form (3.50), we obtain the equality

$$u_0^r(x, \omega_0) = \sum_{k=1}^{N} \varphi_k(\omega_0)\frac{e^{ik_0 R'}}{R'}\text{rect}\left(\frac{x_1}{2a}\right)\text{rect}\left(\frac{x_2}{2a}\right),$$

where $R'^2 = (x_1')^2 + (x_2')^2 + h^2$. Expanding now the function R' into a Taylor series and, analogously, leaving only the first two terms, we obtain instead of (3.51) the approximate representation of solution

$$\tilde{f}^r(x,\omega_0) = \sum_{k=1}^{N} \varphi_k(\omega_0) \int_{-a}^{a}\int_{-a}^{a} \exp\Big\{-ik_0\left[x_1'\left(x_1 - y_1^k\right) + x_2'\left(x_2 - y_2^k\right)\right]$$

$$+ \frac{x_1'^2 + x_2'^2}{2}\left(\frac{1}{h_\varepsilon} - \frac{1}{h}\right)\Big\} dx_1' x_2'. \qquad (3.14)$$

This expression is the mathematical representation of the Fresnel approximation for the field diffracting in a square hole with a lens of focal length $F = h_\varepsilon h/(h_\varepsilon - h)$ (Born and Wolf, 1973; Erokhin, 1978; Goodman, 1976). Approximation (3.52) without including the square terms, or, in other words, for $\varepsilon = 0$ ($h_\varepsilon = h$, $F = \infty$) is called by the Fraunhofer approximation. The image in the Fraunhofer approximation gives sufficiently good approximation for the Fresnel approximation even for aperture dimensions of the same order as the wave length $\lambda_0 = 2\pi c/\omega_0$. Further, for simplicity, we shall consider the Fraunhofer approximation. In this case, instead of 3.25 we shall have

$$\tilde{f}^r(x,\omega_0) = \sum_{k=1}^{N} \varphi_k(\omega_0) \frac{\sin\left[\beta(x_1 - y_1^k)\right]}{x_1 - y_1^k} \frac{\sin\left[\beta(x_2 - y_2^k)\right]}{x_2 - y_2^k}, \qquad (3.15)$$

where $\beta = k_0 a/n$.

Consider, first, the problem of determining the correctness function $\omega(\tau)$ (Lavrent'ev, 1962) of the operator for determining the values of weight coefficients $\varphi_k = \varphi_k(\omega_0)$ in the case when the source coordinates are known.

Formalization of this problem has the following form. Let a certain operator D map the set of generalized functions (3.50), which we denote by L, into the functions $f^r(x,\omega_0)$ of the form (3.51), which we shall denote by F. As it is easy to see, the operator D is the operator of convolution with the kernel

$$K_0(x_1, x_2) = \frac{\sin(\beta x_1)}{x_1} \cdot \frac{\sin(\beta x_2)}{x_2}.$$

The topology in F we define by means of the $L_2(R^2)$ norm. It is well-known that the weak topology in L is, generally speaking, not metrizable. But if $L \in C^1(K)$, where C is the space of continuous functions, K is a compact set in R^2, then we get such a possibility in a certain compact set $M \subset L$, which in this case has the form

$$M = \left\{ f = \sum_{k=1}^{N} \varphi_k \delta(x - y^k) \,\middle|\, 0 < c_1 \leq \varphi_k \leq c_2 < \infty, \quad y^k \in K \right\}. \tag{3.16}$$

On the basis of theorems from Danford and Schwarz (1962), Ivanov *et al.* (1978), there exists a continuous dependence of solution to the inverse problem of determining φ_k, y^k, y_2^k on initial data. In this case, solution stability is characterized by the correctness function

$$\omega(\tau) = \sup_{f, f \in M} \left\{ \rho(f, f') \,\middle|\, \left\| \tilde{f}^r - \tilde{f}^{r'} \right\| \leq \tau \right\},$$

where $\rho(f, f')$ is a certain metric in M. For our case, let

$$\rho(f, f') = \left(\sum_{k=1}^{N} (\varphi_k - \varphi'_k)^2 \right)^{\frac{1}{2}}.$$

It may be shown that the expression (*) is represented as the inequality for a certain square form

$$\sum_{i,j=1}^{N} z_i z_j a_{ij} \leq \tau^2,$$

where

$$z_i = \varphi - \varphi'_i; \quad a_{ij} = \frac{\sin\left[\beta(x_{1i} - x_{1j})\right]}{x_{1i} - x_{1j}} \cdot \frac{\sin\left[\beta(x_{2i} - x_{2j})\right]}{x_{2i} - x_{2j}}.$$

Determining the correctness function is reduced to the well-known problem of determining the maximum of the length z of the radius vector $(z_1, \ldots, z_k)$ connecting the origin with a point in the ellipse

$$\sum_{i,j=1}^{N} z_i z_j a_{ij} = \tau^2/\pi^2.$$

Therefore,

$$\omega(\tau) = \frac{\tau}{\pi\sqrt{\lambda_{\min}}}.$$

Here, $\omega(\tau)$ is the minimal eigenvalue of the symmetric matrix $\{a_{ij}\}$ with diagonal prevalence.

It is clear that $\lambda_{\min} \le a_{ij} = \beta^2$ and the equality holds only if $a_{ij} = 0$, $i \ne j$, i.e., when the ellipse is degenerated into a circle. This may be only if $x_{1i} - x_{1j} = \frac{\pi}{\beta}k$; $x_{2i} - x_{2j} = \frac{\pi}{\beta}l$, where $k = 1, 2, \ldots$; $l = 1, 2, \ldots$. Thus, if the sources are disposed in the knots of a uniform net with the step $\frac{\pi}{\beta}$, determining the weight $\varphi_k(\omega_0)$ has maximal stability, i.e., the correctness function will be minimal

$$\omega(\tau) = \frac{\tau}{\pi\beta} = \frac{\tau h}{2k_0 a}.$$

Hence follows that solution stability of this problem is improved with decreasing of the wave length and of the distance h and, also, for increasing of the size of receiving aperture a.

Suppose for a certain source we know the intensity $\varphi(\omega_0)$, but its coordinates x_1^0, x_2^0 are unknown. Define the correctness function of determining the coordinates.

The correctness set we define in a way analogous to (3.54) for $N = 1$. The compact set M will be the square

$$b_1 \le x_1, x_2 \le b_2,$$

where $(b_2 - b_1) \le 3\pi/2\beta$. The metric in M will be as follows

$$\rho(f, f') = \max\left\{(x_1^0 - x'^0_2), (x_2^0 - x'^0_2)\right\}.$$

For such correctness function it may be shown that (Erokhin, 1979)

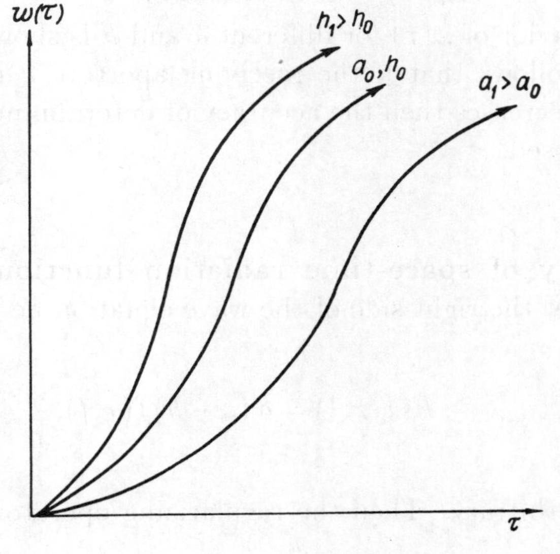

Figure 3.22. The correctness functions for different values of aperture and distance.

$$\omega(\tau) = \sup_{f,f'} \left\{ \rho(f, f') \,\middle|\, \frac{\sin\left[\beta(x_1^0 - x'^0_1)\right]}{x_1^0 - x'^0_1} \right.$$

$$\left. \times \frac{\sin\left[\beta(x_2^0 - x'^0_2)\right]}{x_2^0 - x'^0_2} > \beta - \frac{\tau^2}{2\pi^2\beta c_1^2} \right\}.$$

Hence we fined

$$\omega(\tau) = f^{-1}\left(\beta - \frac{\tau^2}{2\pi^2\beta c_1^2}\right),$$

where $f(x) = \sin(\beta x)/x$, $x \in [0, 3\pi/2\beta]$.

The behavior of $\omega(\tau)$ for different a and h is shown in Figure 3.22. From here follows that if the receiving aperture a increases and the distance h decreases then the accuracy of determining the source coordinates is raised.

Recovery of space-time radiation function for a moving source. Let the right side of the wave equation be as follows

$$F(x, z, t) = \delta'(z - h) f(x, t),$$

where f is arbitrary. Then the regularizing operator has the esimate (3.44).

According to the algorithm it is necessary to calculate the three-dimensional Fourier transforms in sufficiently shallow space - time net. Consider the two-dimensional analogue of this problem, i.e., when the source has the form

$$F(x, z, t) = \delta'(z - h) f(x, t); \quad x, z \in R^1,$$

with the information

$$u_0(x, t) = u(x, z, t)\Big|_{z=0}.$$

First, let

$$f(x, t) = \delta(x - vt)(1 + \cos\omega_0 t), \tag{3.17}$$

where $v/c = 2/3$, $\omega_0 = 8$ Hz, $h = \lambda_0$, $\lambda_0 = 2\pi c/\omega_0$- is the wave length of the radiation typical for the source.

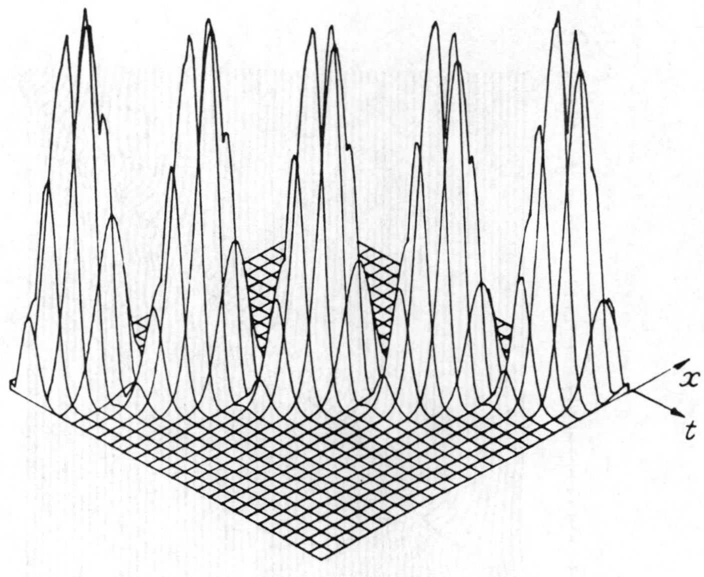

Figure 3.23. The source function $f(x,t)$.

The function $f(x,t)$ is represented in Figure 3.23. The vibration field of this source in the line $z = 0$ will have the form represented in Figure 3.24. In the x-axis we have the coordinates x of the receiving aperture (from $-a$ to a). In the y-axis we have the time $(-T, T)$. Here $a = 2\lambda_0$, $\lambda_0 = 2\pi c/\omega_0$.

Consider the case when the regularization parameter $\Lambda = 0$, i.e., we use only homogeneous waves. Note, that this holds also for $\Lambda > 0$, but only if to initial information we apply the "true filtration" technique (Klayerboat, 1981). In Figure 3.25 we see the recovered function $f(x,t)$ for $\Lambda = 0$. From the comparison of Figure 3.23 and 3.25 we see that even without noise in the information, the right side of (3.39) is not satisfactorily recovered. This is so because for $\Lambda = 0$ we neglect, as is usually done in seismology, the domain $\lambda > |\omega|/c$ in the spectrum of the function $u_0(x,t)$, where for the right side (3.39) we have almost all information for the source (Figure 3.18). Really, the field spectrum of (3.55) has the form

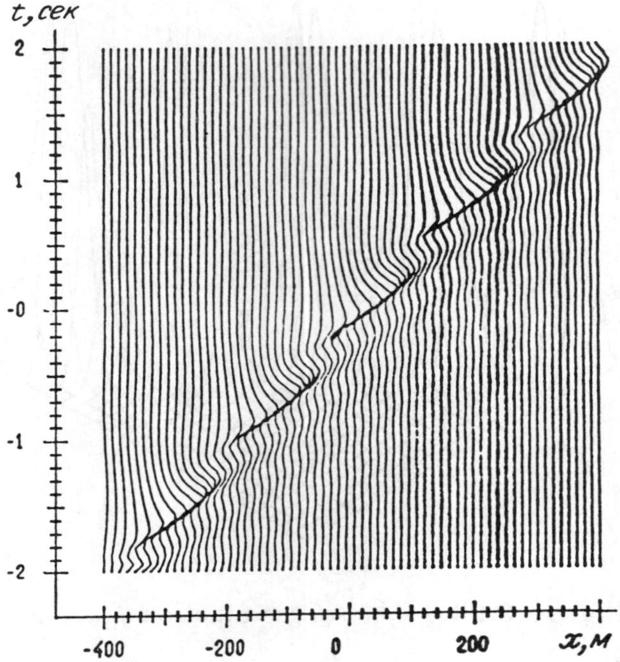

Figure 3.24. Theoretical traces of vibrations of the moving source.

$$\hat{u}_0(\lambda,\omega) = \left(\delta(\omega - \lambda v - \omega_0) + \delta(\omega - \lambda v)\exp(-h\sqrt{\lambda^2 - \left(\frac{\omega}{c} + i\alpha\right)^2})\right),$$

and neglecting the domain $\lambda > |\omega|/c$, we miss the information on the second term.

Now let $\Lambda > 0$. Comparing the functions $f(x,t)$ for $\Lambda = 0,1\Lambda_{\max}$, where $\Lambda_{\max} = \pi/\Delta x$ (Figure 3.26) for $\Lambda = 0$ (Figure 3.25) and the initial function (Figure 3.23), we see that using inhomogeneous waves in recovery of right sides of our equation is of great importance.

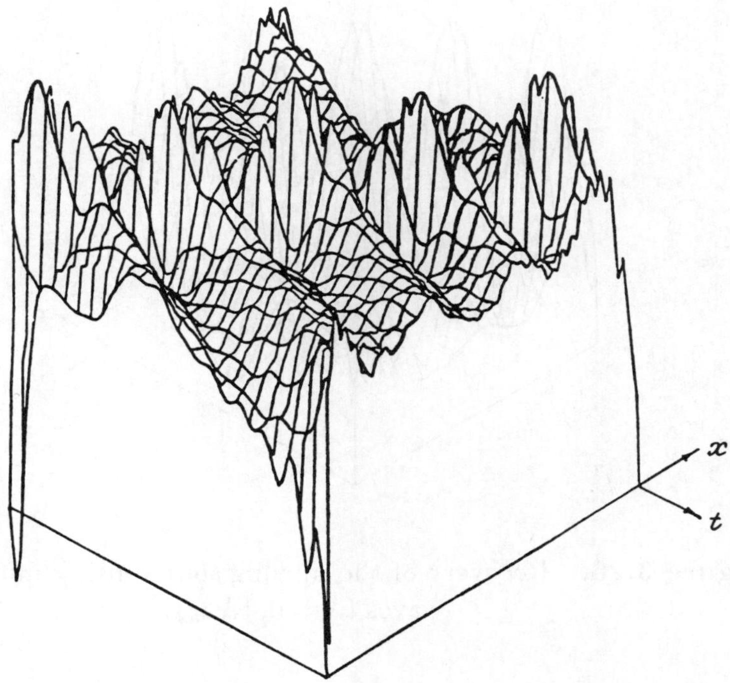

Figure 3.25. Recovery of the moving source using only homogeneous waves ($\Lambda = 0$).

Now consider the regularizing properties of operator (3.42) for the case of noise in data. In Figure 3.27 are represented the theoretical traces of the field u_0^δ if noise is present $||u_0^\delta - u_0|| \leq \delta$. Here the function $f(x,t)$ is also of the form (3.54), $v/c = 2/3$, $h = \lambda_0$, and the relative error in data u_0^δ is 25%.

In Figure 3.28 we have the results of recovery of the function $f(x,t)$ for different Λ. As we may see, for $\Lambda = 0.1\Lambda_{\max}$ the function is recovered the best and when the regularization parameter increases, ($\Lambda = 0.9\Lambda_{\max}$) or decreases ($\Lambda = 0$), the image is worse. Thus, according to obtained theoretical estimates, in each concrete case there exists a centre regularization parameter $\Lambda = \Lambda_{\text{opt}}$ for which the method corresponding to this parameter is the best for solving the inverse problem. The fact that if noise is present in data, then $\Lambda_{\text{opt}} > 0$ verifies the necessity of using inhomogeneous waves in the recovery algorithm.

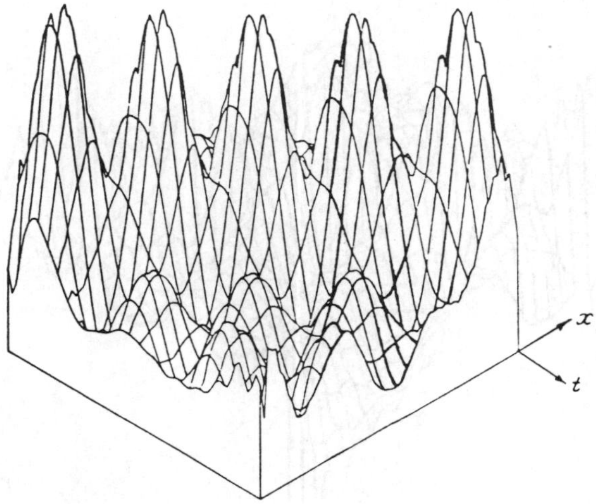

Figure 3.26. Recovery of the moving source using inhomogeneous waves ($\lambda = 0,1\Lambda_{\max}$)

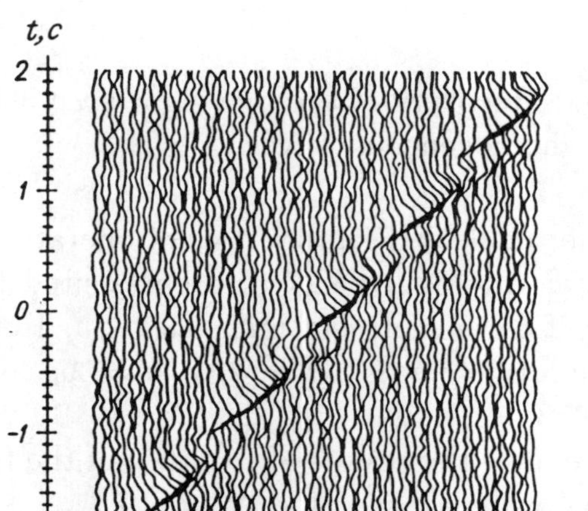

Figure 3.27 Theoretical traces of vibrations of a moving source for noise level of 25%.

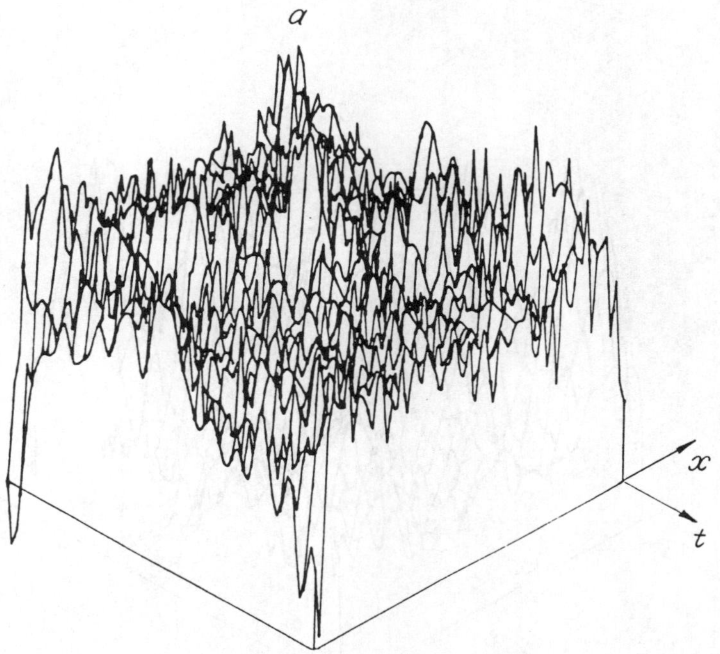

Figure 3.28a. Recovery of the moving source using distorted information, when we use in the algorithm. Homogeneous waves ($\Lambda = 0$).

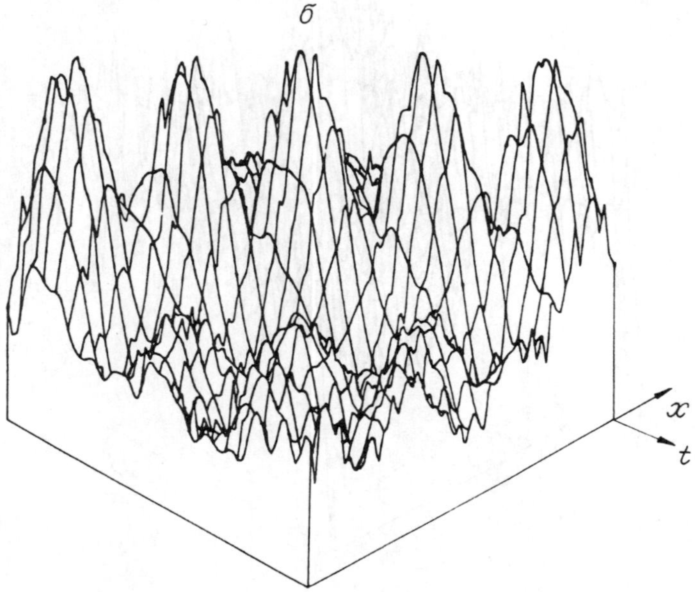

Figure 3.28b. Recovery of the moving source using distorted information, when we use in the algorithm. Inhomogeneous waves ($\Lambda = 0.1\Lambda_{\max}$).

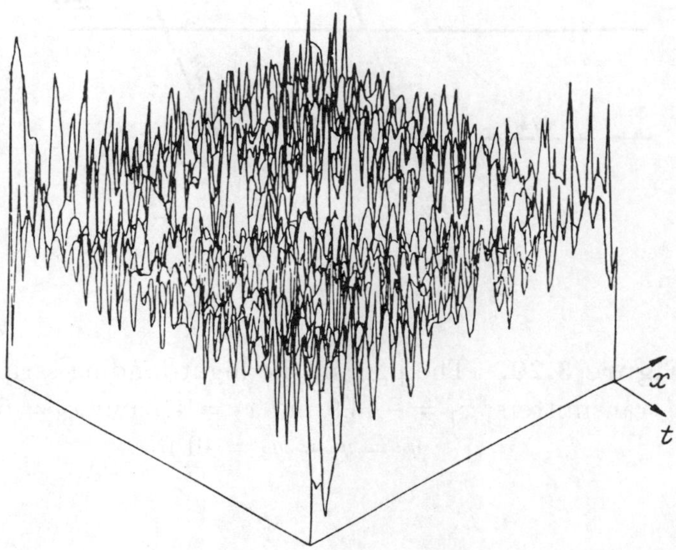

Figure 3.28c. Recovery of the moving source using distorted information, when we use in the algorithm. Inhomogeneous waves ($\Lambda = 0.9\Lambda_{\max}$).

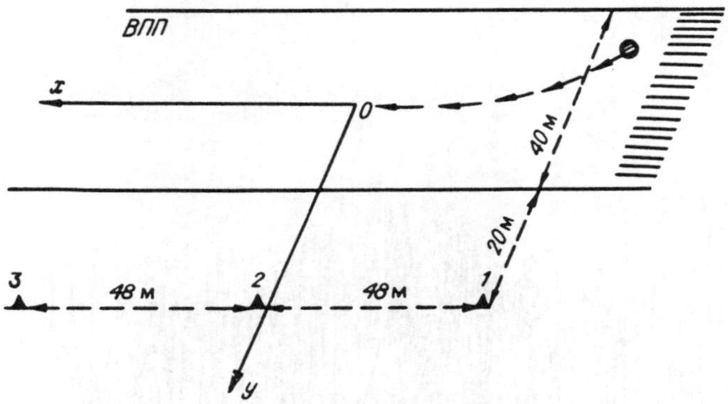

Figure 3.29. The plan of the flight-landing strip. 1-3 the transmitters; $x_1 = -43.9$ m.; $x_2 = 4.1$ m.; $x_3 = 52.1$ m.; $y_1 = y_2 = y_3 = 40$ m.

§3.4 The problem of distance quality control of an airplane landing.

In this paragraph we consider the problem of distance quality control of landing an airplane on a flight-landing strip (FLS) on the basis of analysis of seismic records of the wave process which arises during the landing. For this purpose we give an analytic description of the source which simulates the airplane landing and justify the chosen mathematical model which describes the process of disturbance propagation in the medium.

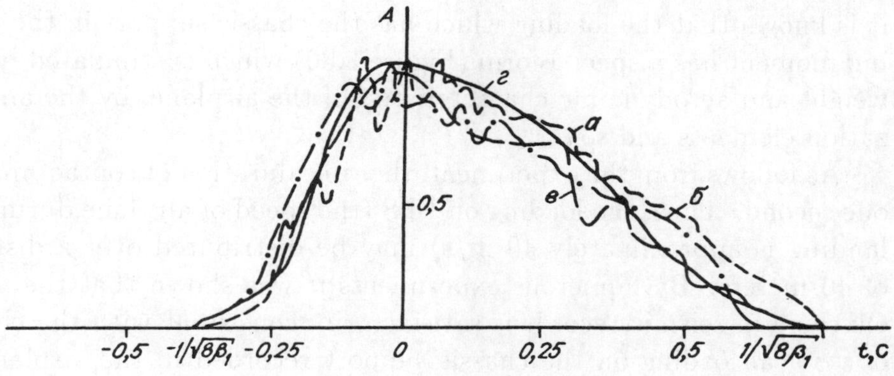

Figure 3.30. The loadings which have the chassis support at the moment of landing (a, b, c) and the theoretical curve used for simulation (d).

For this purpose, at first, in one of our aerodromes, we constructed a system in which was included the transmitters capable of changing the speed of ground shift near the place of probable contact of the airplane and FLS. (Figure 3.29).

In this problem, it is necessary to construct a mathematical model that will characterize the strength of contact of the airplane and the strip, the time duration of contact and the coordinates of the points of the maximal contact strength by the seismic information measured in three points.

Therefore, we shall, at first, describe the mechanism of airplane action onto the FLS.

From the practice of the flight tests (Gudkov and Leshakov, 1968) it is known that the loading which has the chassis support in the landing moment has a special form (Figure 3.30) which is stipulated by the weight and aerodynamic characteristics of the airplane, by the amortization elements and so on.

As follows from the experiment, the time duration of contact may be one second. Thus, the loading on FLS (the speed of airplane during the landing is approximately 40 m/s) may be distributed over a distance of 40 meters. By numerous experiments it was shown that the signal envelope in seismic records is satisfactorily consistent with the change of vertical loading on the chassis support recorded in the airplane by accelerometer. The main frequency of the envelope scaffolding lies in the interval from 2 to 9 Hz. In this case, the comparison of the same signal for different traces (in different points of receiving) (Anikonov *et al.*, 1982) shows the presence of a Doppler frequency shift. This is so because the speed of the signal is close to the speed of the landing airplane. The envelope scaffolding is present because during the ineraction of the chassis wheel with the surface of FLS in the latter periodic disturbances are generated whic occur all along all the line of contact of the airplane with the FLS. The higher the landing speed, the higher is the scaffolding frequency.

It is known that the choice of mathematical model of the physical process is a complicated problem. The suggested model must adequately describe the phenomenon and be sufficiently simple. Therefore, before we chose it, we carried out some experiments to obtain the main peculiarities of the investigated process. They show that the main influence on the seismic record character is the fact that the airplane landing speed is of the same order as the seismic waves speed and measurements are made close to the point of wave excitation. These facts do not allow us to take into account the reflected waves or to separate the solid and surface waves.

In this connection, the wave equation with constant coefficients probably describes the process sufficiently well the main problem, therefore, is to describe a source that simulates the action of the airplane onto the FLS surface.

As the result of experiments, we have chosen the following function describing the loading on the chassis support

$$\varphi(t) = e^{-(\beta_1 t)^2}\Theta(t) + e^{-(\beta_2 t)^2}(1 - \Theta(t)),$$

where $\Theta(t)$ is the Heaviside function. Its graph is represented in Figure 3.30(d). We may note that the theoretical curve approximates the experimental dependences quite well.

For definitiness, the x-axis lies along the FLS and the y-axis lies in the plane of the earth surface; the z-axis will be orthogonal to the earth surface. Then the right side of the wave equation we take in the form

$$F(x, y, z, t) = \delta(z) f(x, y, t).$$

Consider now the case of successive j-multiple contact of the pneumatics of separate chassis supports in the landing process. Denote by x_{kj} the coordinates of touching of the k-wheel with the FLS; v is the airplane speed which is assumed to be constant; c is the propagation speed of elastic waves. Then, for simultaneous touching of the FLS by several chassis supports, represent the source in the form

$$f(x, y, t) = \sum_{k=1}^{3} \delta(x - vt - x_k)\delta(y - y_k) f_k(t),$$

where x_k, y_k are the coordinates of the k-wheel when the mass centre of the airplane is in the origin ($t = 0$). On the basis of these facts, represent the function $f_k(t)$ as follows

$$f_k(t) = \sum_{j=1}^{n_k} A_{kj}\psi_{kj}\left(t - \frac{x_{kj}}{v}\right) e^{i(\omega_{kj}t - \xi_{kj})},$$

where

$$\psi_{kj} = e^{-(\beta_{kj}^{(1)} t)^2}\Theta(t) + e^{-(\beta_{kj}^{(2)} t)^2}(1 - \Theta(t)).$$

Here n_k is the number of contacts by the k-wheel; A_{kj} is the maximal value of loading; ω_{kj} is the frequency of vibrations at jth contact; ξ_{kj} is the phase of vibrations at jth contact.

Further, let α be the absorption coefficient; then the vibration field in the plane $z = 0$ will be as follows

$$u_0(x,y,t) = \sum_{k=1}^{3} \exp\left(-\alpha \frac{\tau_k \gamma}{1-\gamma^2} + \frac{R_k}{\sqrt{1-\gamma^2}}\right) \cdot \frac{1}{R_k} \cdot \sum_{j=1}^{n_k} A_{kj}$$
$$\times\, \psi_{kj}\left(\tilde{t} - \frac{x_{kj}}{v}\right) \exp\left\{i(\omega_{kj}\hat{t} + \xi_{kj})\right\}, \qquad (4.1)$$

where

$$\tilde{t} = t - \frac{\tau_k \gamma}{c(1-\gamma^2)} - \frac{R_k}{c\sqrt{1-\gamma^2}};$$

$$R_k^2 = \frac{\tau_k^2}{1-\gamma^2} + (y - y_k)^2;$$

$$\tau_k = x - vt - x_k; \gamma = \frac{v}{c}.$$

To compare the results of numerical simulation with the experiment, one of the landings was taken where the coordinates of contact, time duration, and loading were fixed by the airplane apparatus. The coordinates of disposition of three seismic transmitters are given in Figure 3.29. The point $t = 0$ coincides with the moment when the mass centre of the airplane is in the origin.

The centre of the first contact had the coordinates $x_{11} = -57$ m; $y_{11} = -5$ m, i.e., was situated before the first transmitter and the contact was done by the right chassis support ($k = 1$). The second contact was done by the left chassis support ($k = 2$). The impact was applied near the second transmitter; its centre had the coordinates $x_{22} = 0$ m, $y_{22} = 5$ m. Finally, the third impact was applied near the third transmitters; its centre had the coordinates $x_{33} = 40$ m, $y_{33} = 0$ m. Further, we shall assume for simplicity that $x_k = 0$.

The durations of contact were determined from readings of devices in the airplane and were, respectively, $\tau_1 = 0.7$ s, $\tau_2 = 1$ s, $\tau_3 = 0.8$ s. The duration and asymmetry were given by means of the parameters $\beta_{kj}^{(1)}$ and $\beta_{kj}^{(2)}$

$$\beta_{11}^{(1)} = \beta_{11}^{(2)} = 3.37; \quad \beta_{22}^{(1)} = 2.9; \quad \beta_{22}^{(2)} = 3.37; \beta_{33}^{(1)} = 2.3; \quad \beta_{33}^{(2)} = 5.06.$$

The amplitudes A_{kk} were taken relatively the second contact - the most powerful:

$$A_{22} = 1; \quad A_{11} = 0.7; \quad A_{33} = 0.8.$$

The damping factor was taken 0.01 (1/m).

In Figure 3.31 are represented the levels of equal shifts normed relative to the maximum. The isolines are given in decibels.

$$gB = 20lg\frac{u(x,y,t)}{u_{\max}(x,y,t)},$$

where $y = 40$ m. The x coordinate is between -100 m and 100 m. The time is between -3 s and 3 s. Thus, in Figure 3.31 are represented the isolines of function (3.56) if we set $\omega_{kj} = \xi_{kj} = 0$. The speed of propagation of seismic waves is equal to 103 m/s and the landing speed of the airplane $v = 43$ m/s. In this Figure we see all three contacts of the airplane with the FLS. The vertical lines are the dispositions of all three transmitters used in the real experiment.

In Figure 3.32 are represented experimental and theoretical seismograms. The overall frequencies used and their phases have the following values

$$\omega_{11} = 6\text{Hz}, \quad \omega_{22} = 3\text{Hz}, \quad \omega_{33} = 4\text{Hz},$$

$$\xi_{11} = 0, \quad \xi_{22} = -\frac{\pi}{2}, \quad \xi_{33} = \pi.$$

As we may see, the model curves approximate quite well the experimental seismograms. Moreover, they not only give the dynamic of the real signal and the kinematic of the process, but regularly describe the typical pecularity - the presence of a Doppler frequency change in the supporting signal.

On the basis of the obtained model, calculations were made of the parameters characterizing the landing. These were found to be of satisfactory accuracy (Anikonov *et al.*, 1982).

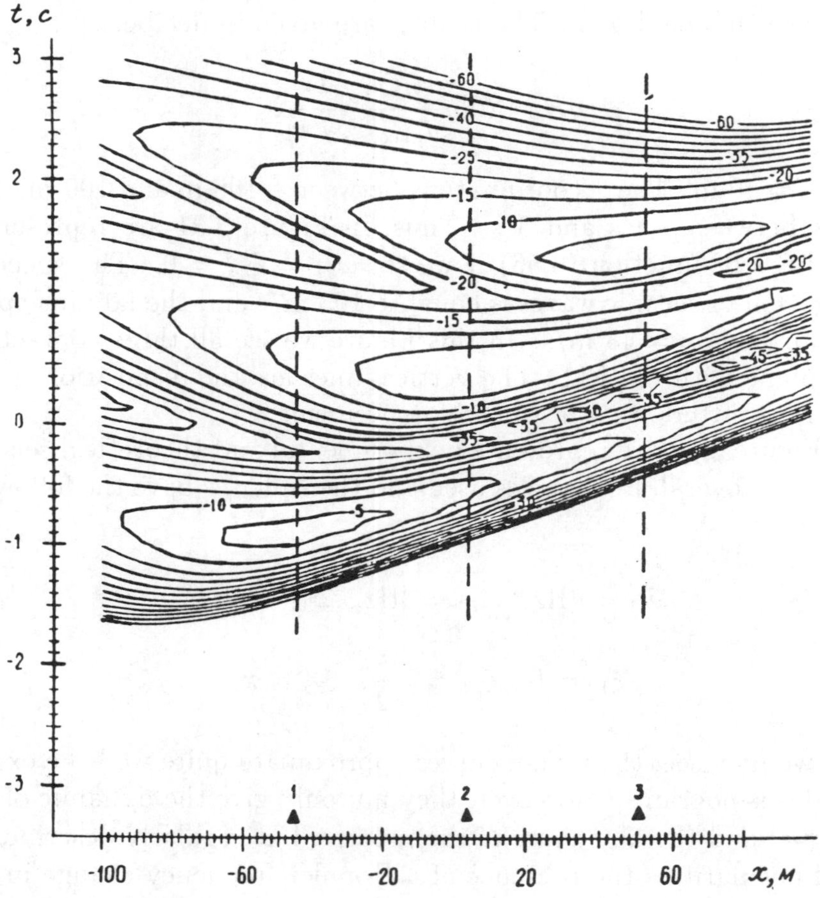

Figure 3.31. The levels of equal amplitudes (db) relative to the maximal value.

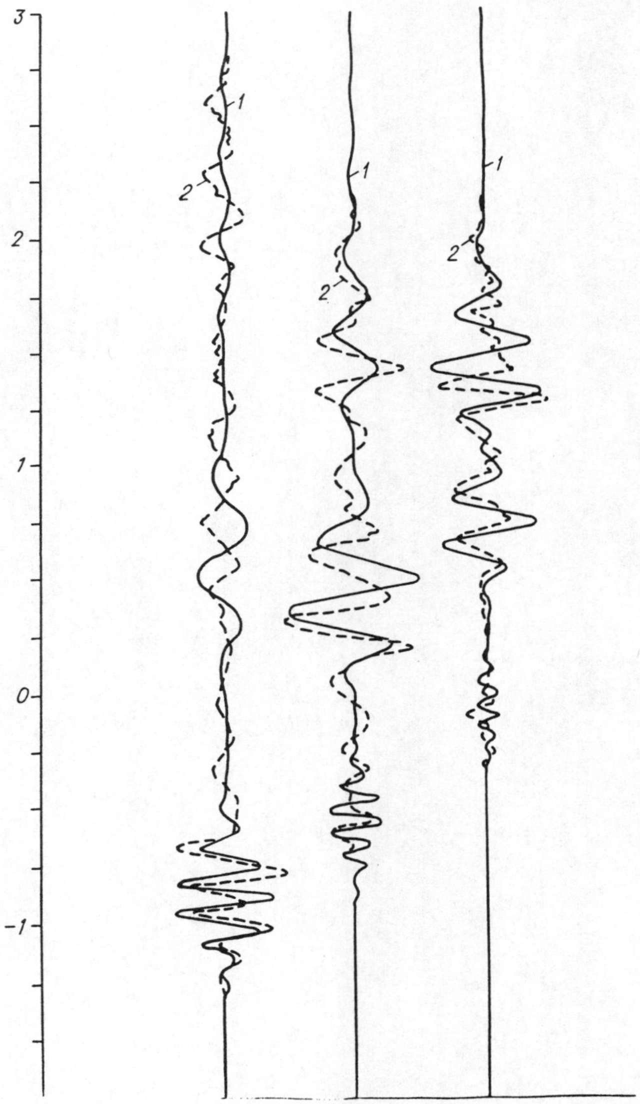

Figure 3.32. The seismograms: 1 - theoretical; 2 - experimental.

Bibliography

Aki, K. and Richards, P. (1983). *Quantitative Seismology. Theory and Methods.* In two volumes. Mir, Moscow (in Russian).

Aleksandrov, A. D. (1937). To the theory of mixed volumes of convex bodies. *Math. Zbornik,* **2**(44), 5, 947-972; 6, 1205-1238 (in Russian).

Aleksandrov, A. D. (1938). To the theory of mixed volumes. *Math. Zbornik,* **3**(45), 1, 27-44, 2, 228-251 (in Russian).

Alekseev, A. S. (1962). Some inverse problems of the theory of wave propagation. *Izv. Akad. Nauk SSSR, Ser. Geof.* **11**, 1514-1531 (in Russian).

Anikonov, Yu. E. (1968). Remarks on convex surfaces. *Sibirsk. Math. J.* **9**, 1, 1413-1415 (in Russian).

Anikonov, Yu. E. (1969). The uniqueness theory for convex surfaces. *Mat. Zametki,* **6**, 1, 115-117 (in Russian).

Anikonov, Yu. E. (1978). *Some Methods for Solving Multidimensional Inverse Problems for Differential Equations.* Sibirsk. Otdel., Nauka, Novosibirsk (in Russian).

Anikonov, Yu. E. (1986). *Pseudodifferential Operators and Inverse Problems.* Preprint, Sib. Otdel. Akad. Nauk SSSR, Vych. Ts., 671 (in Russian).

Anikonov, Yu. E. (1987). New mathematical models and inverse problems. *Mathematical Simulation in Geophysics.* Sib. Otdel. Akad. Nauk SSSR, Vych. Ts., Novosibirsk, 3-42 (in Russian).

Anikonov, Yu. E. (1988). New equations and formulas in inverse problems. *Complex Analysis and Mathematical Physics.* Sib. Otdel. Akad. Nauk SSSR, Vych. Ts., Krasnoyarsk, 20-28 (in Russian).

Anikonov, Yu. E. and Belov, Yu. Ya. (1989). On unique solvability of one inverse problem for the parabolic equation. *Dokl. Akad. Nauk SSSR*, **130**, 6, 1289-1293 (in Russian).

Anikonov, Yu. E. and Bubnov, B. A. (1987). One method of research for multidimensional inverse problems. *Methods of Research for Inverse and Ill-Posed Problems.* Sib. Otdel. Akad. Nauk SSSR, Vych. Ts., Novosibirsk, 22-40 (in Russian).

Anikonov, Yu. E. and Bubnov, B. A. (1988a). Existence and uniqueness of solving inverse problems for the parabolic equation. *Dokl. Akad. Nauk SSSR,* **298**, 4, 777-779 (in Russian).

Anikonov, Yu. E. and Bubnov, B. A. (1988b). Inverse problems for the transfer equation.*Dokl. Akad. Nauk SSSR*, **299**, 5, 497-500 (in Russian).

Anikonov, Yu. E. and Bubnov, B. A. (1989). Control problems and inverse problems. *Dokl. Akad. Nauk SSSR*, **304**, 2, 309-312 (in Russian).

Anikonov, Yu. E. and Erokhin G. N. (1984). Inverse problems of wave theory with dispersion. *Methods for Solving Ill-Posed Mathematical Problems and Problems of Geophysics.* Sib. Otdel. Akad. Nauk SSSR, Novosibirsk, 7-13 (in Russian).

Anikonov, Yu. E. and Golubyatnikov, V. P. (1978). To the uniqueness of solving inverse problems of scattering theory. *Inverse Problems for Differential Equations,* Sib. Otdel. Akad. Nauk SSSR, Vych. Ts., Novosibirsk, 13-17 (in Russian).

Anikonov, Yu. E. and Marchuk, A. G. (1975). To the inverse diffraction problem. *Mat. Probl. Geof.* Sib. Otdel. Akad. Nauk SSSR, Vych. Ts., Novosibirsk, 54-62 (in Russian).

Anikonov, Yu. E. and Stepanov, V. N. (1981). Uniqueness and stability of solving one problem of geometry in whole. *Mat. Zbornik,* **116**, 12, 539-546 (in Russian).

Anikonov, Yu. E., Borisov, B. T., and Erokhin, G. N. (1982). *Simulation of One Wave Equation.*Preprint, Sib. Otdel. Akad. Nauk SSSR, Vych. Ts., 387 (in Russian).

Ayupova, N. B. (1986). Formulas in the problems of source determining. *Correctness Problems and the Research Methods for Solving Inverse Problems.* Sib. Otdel. Akad. Nauk SSSR, Vych. Ts., Novosibirsk, 47-57 (in Russian).

Bakelman, I. Yu., Verner, A. L., and Kanter, B. E. (1973).*Introduction to Differential Geometry in Whole.* Nauka, Moscow (in Russian).

Bejtman, G. and Erdein, A. (1969). *Tables of Integral Transformations.* **1**. Nauka, Moscow (in Russian).

Blokhintsev, D. I. (1981). *Acoustics of Inhomogeneous Moving Medium.* Nauka, Moscow (in Russian).

Blyashke, V. (1967). *Disk and Ball.* Nauka, Moscow (in Russian).

Boonnesen, T. and Fenchel, W. (1978). *Theorie der Konvexen Kornor.* New-York.

Born, M. and Wolf, E. (1973). *The Basis of Optics.* Nauka, Moscow (in Russian).

Brychkov, Yu. A. and Prudnikov, A. P. (1977). *Integral Transformations of Generalized Functions.* Nauka, Moscow (in Russian).

Bubnov, B. A. (1986). Existence and uniqueness of inverse problem solutions for parabolic and elliptic equations of mathematical physics. *Nonclassical Equations of Mathematical Physics.* Sib. Otdel. Akad. Nauk SSSR, Vych. Ts., Novosibirsk, 226-235 (in Russian).

Bubnov, B. A. (1987a). Solvability of multidimensional inverse problems for the telegraph equation. *Uniqueness, Stability, and Methods for Solving the Mathematical Physics Problems.* Sib. Otdel. Akad. Nauk SSSR, Vych. Ts., Novosibirsk, 33-41 (in Russian).

Bubnov, B. A. (1987b). *To the Solvability of Multidimensional Inverse Problems for Parabolic Equations.* Preprint. Sib. Otdel. Akad. Nauk SSSR, Vych. Ts., Novosibirsk, 714 (in Russian).

Bubnov, B. A. (1987c). *To the Solvability of Multidimensional Inverse Problems for Hyperbolic Equations.* Preprint. Sib. Otdel. Akad. Nauk SSSR, Vych. Ts., Novosibirsk, 713 (in Russian).

Bubnov, B. A. (1987d). New settings of multidimensional inverse problems for hyperbolic equations in the classes of finite smoothness functions. *Applying the Methods of Functional Analysis in Mathematical Physics Equations.* Sib. Otdel. Akad. Nauk SSSR, Inst. of Math., Novosibirsk, 173-185 (in Russian).

Bubnov, B. A. (1988). Necessary and sufficient conditions of solvability of inverse problem for the hyperbolic equation in the class of t-periodic functions. *Applications of Functional Analysis Methods to Nonclassical Mathematical Physics Equations.* Sib. Otdel. Akad. Nauk SSSR, Inst. of Math., Novosibirsk, 174-185 (in Russian).

Bukhgeim, A. L. (1983). *Volterra Equations and Inverse Problems.* Nauka, Moscow (in Russian).

Butkovskii, A. G. (1979). *Characteristics of the Systems with Distributed Parameters.* Nauka, Moscow (in Russian).

Butkovskii, A. G. (1985). *Methods and Control Systems for the Systems with Distributed Parameters.* Nauka, Moscow (in Russian).

Danford, N. and Schwarz, Dj. (1962). *Linear Operators.* **1**, Nauka, Moscow (in Russian).

Dezin, A. L. (1980). *General Questions of Boundary Problems Theory.* Nauka, Moscow (in Russian).

Ditkin, V. A. and Prudnikov, A. P. (1974). *Integral Transformations and Operational Calculus.* Nauka, Moscow (in Russian).

Erokhin, G. N. (1978). To the problem of diffraction of a scalar field in the hole. *Conditionally Correct Problems of Mathematical Physics and Interpreting Geophysical Observations.* Sib. Otdel. Akad. Nauk SSSR, Vych. Ts., Novosibirsk, 64-74 (in Russian).

Erokhin, G. N. (1979). To the problem of stability of determining the radiating point objects. *Mathematical Methods of Interpreting Geophysical Observations.* Sib. Otdel. Akad. Nauk SSSR, Vych. Ts., Novosibirsk, 35-49 (in Russian).

Erokhin, G. N. (1982). *The Problems of Determining the Right Side of the Wave Equation.* Preprint. Sib. Otdel. Akad. Nauk SSSR, Vych. Ts., Novosibirsk, 348 (in Russian).

Erokhin, G. N. (1984). Elements of structure analysis and inverse problems. *Methods for Solving Ill-Posed Mathematical Problems and Problems of Geophysics.* Sib. Otdel. Akad. Nauk SSSR, Vych. Ts., Novosibirsk, 34-51 (in Russian).

Erokhin, G. N. and Bortnikov, P. B. (1987). Inverse problem of determining the seismic moment tensor of the earthquake centre. *Geology and Geophysics.* **4**, 115-123 (in Russian).

Erokhin, G. N. and Bortnikov, P. B. (1988). Determining the dynamic characteristics of weak earthquake centre on the basis of the seismic moment tensor. *Ibid,* **8**, 120-131 (in Russian).

Gol'tsman, F. M. (1982). *Physical Experiment and Statistic Conclusions.* Leningr. Gos. Univ., Leningrad (in Russian).

Golubyatnikov, V. P. (1982). On recovery of the body's shape and its projections. *Dokl. Akad. Nauk SSSR*, **262**, 3, 521-522 (in Russian).

Gradshteyn, I. S. and Ryzhik, I. M. (1971). *Tables of Integrals, Summs, Serieses, and Products.* Nauka, Moscow (in Russian).

Gudkov, A. I. and Leshakov, P. S. (1968). *External Loadings and Strength of Flying Apparatuses,* Mashinostroenie, Moscow (in Russian).

Ivanov, V. K., Vasin, V. V., and Tanana, V. P. (1978). *The Theory of Linear Ill-Posed Problems and its Applications.* Nauka, Moscow (in Russian).

Kemelkov, D. U. (1989). Solvability of an inverse problem for the Lamé system. *Boundary-Value Problems for Nonclassical Equations of Mathematical Physics.* Sib. Otdel. Akad. Nauk SSSR, Inst. of Math., Novosibirsk, 119-122 (in Russian).

Klayerboat, D. F. (1981). *Theoretical Basis of Geophysical Information Processing.* Nedra, Moscow (in Russian).

Korlyuch, V. S. (1985). *Handbook in Probability Theory and Mathematical Statistics* Nauka, Moscow (in Russian).

Lavrent'ev, M. M. (1962). *On Certain Ill-Posed Problems of Mathematical Physics.* Sib. Otdel. Akad. Nauk SSSR, Nauka, Novosibirsk (in Russian).

Lavrent'ev, M. M. (1972). Inverse problems and special operator equations of the first kind. *International Congress of Mathematicians (Nice, 1970), Papers of Soviet Mathematicians,* Nauka, Moscow, 130-136 (in Russian).

Lavrent'ev, M. M., Romanov, V. G., and Shishatskii, S. P. (1980). *Ill-Posed Problems of Mathematical Physics and Analysis.* Nauka, Moscow (in Russian).

Leibovich, S. and Sibass, A. (1977). *Nonlinear Waves* Mir, Moscow (in Russian).

Lions, J. L. (1973). On optimization of distributed systems control. *Uspekhi Math. Nauk,* **28** , 4, 15-46 (in Russian).

Marchuk, G. I. (1976). Surrounding medium and problems of enterprises displacement. *Dokl. Akad. Nauk SSSR*, **227** , 186, 5, 1056-1059 (in Russian).

Marchuk, A. G. (1978). *Optimal Regularizing Operators and Certain Diffraction Problems.* Kand. Dissert., Sib. Otdel. Akad. Nauk SSSR, Vych. Ts., Novosibirsk (in Russian).

Marchuk, G. I., Agoshkov, V. I., and Shutyaev, V. P. (1986). Conjugated equations and disturbances algorithms in application problems. *Calculating Processes and Systems,* **4**, Nauka, Moscow, 5-62 (in Russian).

Marri, J. (1983). *Nonlinear Differential Equations in Biology.* Mir, Moscow (in Russian).

Mathematical Simulation. Processes in Nonlinear Media (1986). Nauka, Moscow (in Russian).

Petrashen', G. I. (1978). *Problems of Dynamic Theory of Seismic Waves Propagation,* XVIII, *The basis of Mathematical Theory of Elastic Waves Propagation.* Nauka, Leningrad (in Russian).

Petrovskii, I. G. (1938). *Bulletin of Mosk. Gos. Univ.,* **1**, 8, 1-72 (in Russian).

Pogorelov, A. V. (1969). *External Geometry of Convex Surfaces.* Nauka, Moscow (in Russian).

Pogorelov, A. V. (1973). *The Fourth Gilbert Problem.* Nauka, Moscow (in Russian).

Pogorelov, A. V. (1975). *Multidimensional Minkovskii Problem.* Nauka, Moscow (in Russian).

Pshenichny, B. I. and Danilin, Yu. N. (1975). *Numerical Methods in Extremal Problems,* Nauka, Moscow (in Russian).

Rabiner, L. and Gould, B. (1978). *Theory and Application of Digital Processing of Signals.* Mir, Moscow (in Russian).

Rays, J. (1982). *Mechanics of Earthquake centre.* Mir, Moscow (in Russian).

Romanov, V. G. (1980). *Inverse Problems of Mathemtical Physics.* Nauka, Moscow (in Russian).

Romanov, V. G. and Pukhnacheva, T. P. (1987). On unique solvability of determining conductivity of inhomogeneous media. *Mathematical Analysis and Differential Equations.* Sib. Otdel. Akad. Nauk SSSR, Inst. of Math., Novosibirsk 111-120 (in Russian).

Smoes, M. (1984). Chemical waves in the Zhabotinskii vibration system. *Synergetics.* Mir, Moscow, 139-163 (in Russian).

Sobolev, S. L. (1974). *Introduction in the Theory of Cubature Formulas.* Nauka, Moscow (in Russian).

Stefano Campi. (1986). Reconstructing a convex surface from certain measurements of its projections. *Bulletino Union Math. Italy* (6) 5-B, 945-959.

Stepanov, V. N. (1981). Some problems of convex surfaces geometry. *Investigation of Ill-Posed Inverse Problems and Some Operator Equations.* Sib. Otdel. Akad. Nauk SSSR, Vych. Ts., Novosibirsk, 65-75 (in Russian).

Stepanov, V. N. (1984). Geometric questions in the inverse scattering problem. *Sibirs. Math. J.,* **15**, 1, 133-145 (in Russian).

Synergetics. (1984), Mir, Moscow (in Russian).

Tikhonov, A. N. and Arsenin, V. Ya. (1979). *Methods for Solving Inverse Problems.* Nauka, Moscow (in Russian).

Tikhonov, A. N., Goncharskii, A. V., and Stepanov, V. V. (1983). *Regularizing Algorithms and A Priori Information,* Nauka, Moscow (in Russian).

Timoshin, Yu. V. (1978). *Impulse Seismic Holography,* Nedra, Moscow (in Russian).

Trev, F. (1984). *Introduction in Theory of Pseudodifferential Operators. 1. Pseudodifferential Operators,* Nauka, Moscow (in Russian).

Tsibulchik, G. M. (1975). On forming a seismic image on the basis of homography principle. *Geology and Geophysics.* **11**, 97-107 (in Russian).

Vasiljev, F. P. (1981). *Methods for Solving Extremal Problems,* Nauka, Moscow (in Russian).

Vinogradov, M. B., Rudenko, O. V., and Sukhorukov, A. P. (1979). *Waves Theory,* Nauka, Moscow (in Russian).

Vladimirov, V. S. (1976). *Generalized Functions in Mathematical Physics,* Nauka, Moscow (in Russian).

Volkov, Yu. A. (1963). Stability of solution to the Minkovskii problem. *Vestnik LGU,* **1**, 33-43 (in Russian).

Whizem, J. (1977). *Linear and Nonlinear Waves,* Mir, Moscow (in Russian).

Zemanyan, A. G. (1974). *Integral Transformations of Generalized Functions,* Nauka, Moscow (in Russian).